Essential College Pre-algebra

Rachel Sturm-Beiss
Kingsborough Community College

Joshua Yarmish
Pace University

Cover image © Shutterstock, Inc.

www.kendallhunt.com
Send all inquiries to:
4050 Westmark Drive
Dubuque, IA 52004-1840

Chapter			Page

Preface

This textbook is intended for a one semester college course in pre-algebra. By doing so, this workbook is also intended to help prepare students for success in their future algebra/mathematics courses. All topics and concepts in the workbook are presented and organized in a format that is easy for students to follow and convenient for instructors to use as a resource for lesson planning and problem selection. The text prepares the student for an end of semester pre-algebra exam through explanations and worked out examples followed by extensive exercises and worksheets. Answers are provided for all exercises. Problem sets intended for final exam review are included in the Appendix.

Rachel Sturm-Beiss
Joshua Yarmish

Acknowledgements

We would like to thank our colleagues Igor Melamed, and Douglas Henderson for their interest in this project and for their very helpful suggestions.

CHAPTER 1: Whole Numbers

1A – Word names, Addition and Subtraction of Whole Numbers

The whole numbers are as follows $0, 1, 2, 3, 4, 5, 6, 7, ...$

EXAMPLE 1:

A)   Write the word name for 254,789,123 .
B)   Write the number: fifty-three thousand, two hundred forty five.

SOLUTION:

A)   Two hundred fifty- four million, seven hundred eighty- nine thousand, one hundred twenty- three.
B)   53,245

EXAMPLE 2: Add $549 + 357$

SOLUTION:

$$
\begin{array}{r}
{\scriptstyle 1\ 1} \\
549 \\
357 \\
\hline
906
\end{array}
$$

EXAMPLE 3: **A)** Subtract $2067 - 1348$ **B)** Subtract $2008 - 123$

SOLUTION:

A)

$$
\begin{array}{r}
{\scriptstyle 1\ \ 1\ \ 5\ \ 1} \\
2\ 0\ 6\ 7 \\
1\ 3\ 4\ 8 \\
\hline
7\ 1\ 9
\end{array}
$$

B)

$$
\begin{array}{r}
{\scriptstyle 1\ \ 9\ \ 1} \\
2\ 0\ 0\ 8 \\
1\ 2\ 3 \\
\hline
1\ 8\ 8\ 5
\end{array}
$$

EXAMPLE 4:

A)   Ron's monthly apartment expenses are a mortgage payment of $2378, and a maintenance fee of $678. Find his total apartment expenses.

B)   Kim earned $452 in May and $378 in June. How much more did she earn in May than in June?

C)   A garden is 324 feet long and 256 feet wide. How many feet of fencing are needed to enclose the garden?

D) Jake's flight from New York to London departs at 2:45PM and is in the air for 5 hours and 35 minutes. What is the arrival time?

E) John usually finishes work at 5:15PM. On Fridays he ends 2 hours and 30 minutes earlier. What time does he get off work on Fridays?

SOLUTION:

A) $2378 + 678 = \$3056$

B) $452 - 378 = \$74$

C)

```
        324
  ┌──────────────┐
  │              │
256│              │256
  │              │
  └──────────────┘
        324
```

Sum the four sides:

$$324 + 256 + 324 + 256 = 1160 \text{ feet}$$

Or, use the formula for the **perimeter**, $P = 2L + 2W$, where L is the length and W is the width. In this case, $P = 2(324) + 2(256) = 1160$ feet.

D) Line up the times vertically, add the hours and the minutes:

```
  2 : 4 5
  5 : 3 5
 ─────────
  7 : 8 0  =
```

7 hours + 60 minutes + 20 minutes = 7 hours + 1 hour + 20 minutes = 8:20PM

E) Line up the times vertically and subtract. Since the 15 minutes in 5:15 is less than the 30 minutes, borrow 60 minutes from the 5 hours.

```
  4 : 7 5  ← 15 + 60 = 75
  5̶ : 1̶ 5̶
  2 : 3 0
 ─────────
  2 : 4 5  PM
```

1A – EXERCISES

For 1 – 6, write the word name of the whole number.

1. 2,367   **2.** 156,236   **3.** 45,689
4. 57,590,123   **5.** 78,954   **6.** 345,298,240

For 7 - 12, write the number.

7. Fifty-two thousand, two hundred seventy-five

8. One hundred twenty-five thousand, three hundred and five

9. Two thousand, seven hundred fifty

10. Two million, three hundred and five thousand, four hundred eighty-two

11. Twenty-five million, two hundred forty-six thousand, five hundred forty-one
12. Three hundred eighty-two million, four hundred and six thousand, four hundred

For 13 - 30, perform the indicated operation.

13. $189 + 357$	14. $345 + 578$	15. $1890 + 7534$	16. $2356 + 765$
17. $456 + 5689$	18. $674 + 589$	19. $843 + 583$	20. $8732 + 2789$
21. $1456 - 273$	22. $2609 - 1758$	23. $3007 - 2359$	24. $257 - 172$
25. $1270 - 583$	26. $5602 - 2751$	27. $3489 - 1573$	28. $2034 - 1352$
29. $385 + 752 + 25$	30. $362 + 34 + 78$		

31. In January Sam earned $578 in commissions and in February he earned $637. How much did he earn in both months?

32. Arnie spent $457 for a flight to California, $45 for travel insurance and $167 for a hotel. Find the total .

33. Sue bought school books for $56, $165, and $78 . How much did she spend on books?

34. In June a store sold 456 air conditioners and in July they sold 502. How many more air conditioners were sold in July than in June?

35. A University has 5834 arts majors and 1598 science majors. How many more arts than science majors are there?

36. A room is 736 feet long and 365 feet wide. How many feet of base board are required to surround the room?

37. A school yard is 739 feet long and 390 feet wide. How many feet of fencing are required to enclose the school yard?

38. Tom's class begins at 3 : 40PM, and ends 2 hours and 55 minutes later. What time does it end?

39. A movie that runs 3 hours and 25 minutes begins at 7:42 PM. What time does it end?

40. A class that lasts 2 hours and 35 minutes begins at 8 : 25 AM. What time does it end?

41. A class that lasts 2 hours and 25 minutes ends at 4:15 PM. What time does it begin?

42. It takes 2 hours and 25 minutes to fly from New York to Chicago. If the flight arrives at 4:12PM, what time did it depart?

43. It takes 3 hours and 20 minutes to drive from New York City to Baltimore. If a car arrives in Baltimore at 7:15PM, what time did it leave New York City?

1A – WORKSHEET: Word names, Addition and Subtraction of Whole Numbers

For 1 – 3, write the word name of the whole number.

1. 3,765	
2. 156,234	
3. 34,405,002	

For 4 - 9, write the number.

4. Forty-five thousand, three hundred thirty-five
5. Five hundred sixty-two thousand, four hundred and six
6. Seven thousand, five hundred eight-six
7. Five million, four hundred and seven thousand, six hundred forty-two
8. Thirty-six million, three hundred forty-seven thousand, seven hundred forty-one
9. Fifty-seven million, three hundred and four thousand, forty-two

For 10 - 15, Perform the indicated operation.

10. $356 + 782$	**11.** $2356 + 1678$
12. $3567 + 2458 + 534$	**13.** $574 - 298$
14. $5042 - 3651$	**15.** $2003 - 967$
16. Becky sold two sewing machines on ebay for \$458 and \$376. What was her total?	

17.	Jake spent $178, $65, and $189 on text books. How much did he spend in total?

18.	Kim's first rental apartment cost $1278 a month. Her second apartment's rent was $2025 a month. How much more expensive was the second apartment?

19.	A grocery store spends $3459 a month on refrigeration. After an expansion their refrigeration cost is $6735 each month. How much more did they spend on refrigeration after the expansion?

20.	A court yard is 671 feet long and 457 feet wide. How many feet of fencing are required to enclose the court yard?

21.	A flight departs at 3:40 PM. The duration of the flight is 4 hours and 52 minutes. When does the flight land?

22.	It takes 4 hours and 15 minutes to drive from New York City to Boston. If a bus leaves Penn Station for Boston at 5:53 PM, what time should it arrive?

23.	A class meets for 2 hours and 25 minutes. If the class ends at 4:15 PM, what time does it start?

24.	It takes 4 hours and 34 minutes to take a tour boat around Manhattan. If the tour ends at 8:15PM, what time did it start?

Answers:

1. Three thousand, seven hundred sixty-five
2. One hundred fifty –six thousand, two hundred thirty-four
3. Thirty-four million, four hundred and five thousand, two

4.	45,335	5.	562,406	6.	7,586	7.	5,407,642	8.	36,347,741	9.	57,304,042
10.	1138	11.	4034	12.	6559	13.	276	14.	1391	15.	1036
16.	$834	17.	$432	18.	$747	19.	$3276	20.	2256 ft.	21.	8:32 PM
22.	10:08 PM	23.	1:50 PM	24.	3:41 PM						

1B – Multiplication and Division of Whole Numbers

We multiply **factors** and the result is the **product** of the factors. For example, in the equation $2 \times 3 = 6$, 2 and 3 are factors and 6 is the product.

EXAMPLE 1: Multiply: **A)** 345×6 **B)** 728×45

SOLUTION:

A)
```
    2  3
   3  4  5
         6
  ─────────
  2  0  7  0
```

B)
```
        1   3
        1   4
        7  2  8
           4  5
     ───────────
     3  6  4  0
   2  9  1  2
   ───────────
   3  2  7  6  0
```

We divide a **divisor** into a **dividend** and the result is a **quotient**. For example $12 \div 3 = 4$, where 12 is the dividend, 3 is the divisor, and 4 is the quotient.

EXAMPLE 2:   Divide **A)** $1035 \div 23$ **B)** $18312 \div 42$ **C)** $7532 \div 32$

SOLUTION:

A)
```
          4  5
     ┌──────────
  2 3│ 1  0  3  5
       9  2  ↓
     ──────────
       1  1  5
       1  1  5
     ──────────
             0
```

B)
```
          4  3  6
     ┌──────────────
  4 2│ 1  8  3  1  2
       1  6  8  ↓  ↓
     ──────────────
       1  5  1  ↓
       1  2  6  ↓
     ──────────────
          2  5  2
          2  5  2
     ──────────────
                0
```

C)
```
          2  3  5
     ┌──────────────
  3 2│ 7  5  3  2
       6  4  ↓  ↓
     ──────────────
       1  1  3  ↓
          9  6  ↓
     ──────────────
          1  7  2
          1  6  0
     ──────────────
          1  2  ←  remainder
```

EXAMPLE 3:

A)   A room is 67 feet long and 23 feet wide. Find the area of the room.
B)   If one acre of land costs $378, how much do 57 acres cost?
C)   If 75 yards of fabric cost $3975, how much does one yard cost?
D)   If three bottles of wine cost $48 how much do five bottles cost?
E)   Find the cost of carpeting a room that is 25 by 15 feet, if carpeting costs $12 a square foot.
F)   Find the average of the following grades: 93, 85, and 92.
G)   Find the average of the following scores: 25, 36, 44, 38, and 42.

SOLUTION:

A) The area of a room is given by $area = length \times width$ square units, so, area = $67 \times 23 = 1541$ square feet.

B) $378 \times 57 = \$21,546$

C) $3975 \div 75 = \$53$

D) First calculate how much one bottle costs, then multiply by 5.
$48 \div 3 = \$16$, and $16 \times 5 = \$80$

E) First calculate the area of the room (in square feet), then multiply by the cost per square foot.
$area = 25 \times 15 = 375$ square feet
$cost\ of\ carpeting = 375 \times 12 = \$4500$

F) Find the sum of the three grades, then divide by 3.
$average = \frac{93+85+92}{3} = 90$

G) $Average = \frac{25+36+44+38+42}{5} = 37$

Unit conversions and geometry formulas:

Unit conversions:
- 1 foot = 12 inches
- 1 yard = 3 feet = 36 inches
- 1 hour = 60 minutes
- 1 pound = 16 ounces
- 1 gallon = 4 quarts = 16 cups
- 1 quart = 4 cups
- 1 pint = 2 cups

Geometry formulas:
- Perimeter of a rectangle = $2(length) + 2(width)$
- Area of a rectangle = $length \times width$

1B – EXERCISES

For 1 – 18, perform the indicated operation.

1.	353×286		**2.**	483×75		**3.**	1328×56
4.	5342×2005		**5.**	9876×2500		**6.**	487×307
7.	325×207		**8.**	406×354		**9.**	3600×2800
10.	$6396 \div 52$		**11.**	$5842 \div 23$		**12.**	$18864 \div 36$
13.	$26752 \div 32$		**14.**	$63135 \div 207$		**15.**	$46092 \div 23$
16.	$31941 \div 63$		**17.**	$16219 \div 23$ (find quotient and remainder)		**18.**	$45028 \div 56$ (find quotient and remainder)

19. A room is 28 feet long and 15 feet wide. Find the area of the room.

20. A court yard is 205 feet wide and 306 feet long. Find the area of the court yard.

21. If one acre of land costs $4605, how much do 36 acres cost?

22. If 45 yards of a fabric cost $1260, how much does one yard cost?

23. If 75 bags of grass seed cost $3150, how much does one bag cost?

24. If 17 bottles of juice cost $68, how much do 5 bottles cost?

25. If 49 pounds of grapes cost $147, how much do 15 pounds cost?

26. Find the cost of carpeting a room that is 12 by 23 feet, if carpeting is $16 a square foot.

27. Find the cost of tiling a floor if the floor is 30 feet by 24 feet and tiles cost $3 per square foot.

28. Find the average of the following grades: 78, 85, and 92.

29. Find the average of the following basketball scores: 59, 103, and 45.

30. Find the average of the following grades: 75, 83, 95, and 87.

31. Find the average of the following measurements: 12, 15, 23, 29, and 16.

32. 275 inches is equal to : _____ feet and _____ inches.

33. 572 inches is equal to : _____ feet and _____ inches.

34. 103 gallons is equal to: _____ quarts.

35. 203 quarts is equal to: _____ gallons and _____ quarts.

36. 75 ounces is _____ pounds and _____ ounces.

1B – WORKSHEET: Multiplication and Division of Whole Numbers

For 1 – 12, perform the indicated operation.

1. 372×52	**2.** 257×68	**3.** 5782×49
4. 356×405	**5.** 257×350	**6.** 4800×3600
7. $2622 \div 38$	**8.** $3150 \div 42$	**9.** $30315 \div 43$
10. $69161 \div 23$	**11.** $170208 \div 34$ (find quotient and remainder)	**12.** $44928 \div 74$ (find quotient and remainder)

13.	A room is 54 feet long and 27 feet wide. Find the area of the room.
14.	The cost of storage in a warehouse is $56 per square yard. How much do 28 square yards of storage cost?
15.	If an apartment of size 75 square yards is sold for $52,875, what is the cost per square yard?
16.	If a 50 square yard apartment is rented at $1750 per month, how much is the rent of a 75 square yard apartment if the rent is the same per yard?
17.	Find the cost of carpeting a room that is 15 by 28 feet, if carpeting is $12 a square foot.
18.	Find the average of the following grades: 76, 85, 92, and 87.
19.	542 inches is equal to _____ feet and _____ inches.
20.	280 inches is equal to _____ feet and _____ inches.
21.	230 gallons is _____ quarts.

22. 315 quarts is _____ gallons and _____ quarts.

23. 250 ounces is _____ pounds and _____ ounces.

Answers:

1.	19344	**2.**	17476	**3.**	283318	**4.**	144180
5.	89950	**6.**	17280000	**7.**	69	**8.**	75
9.	705	**10.**	3007	**11.**	5006 rem 4	**12.**	607 rem 10
13.	1458 ft^2	**14.**	$1568	**15.**	$705	**16.**	$2625
17.	$5040	**18.**	85	**19.**	45 ft., 2 in.	**20.**	23 ft., 4 in.
21.	920 qts.	**22.**	78 gal., 3 quarts	**23.**	15 lbs., 10 oz.		

1 – Answers to Exercises

Section A

1. Two thousand, three hundred sixty-seven
2. One hundred fifty-six thousand, two hundred thirty-six
3. Forty-five thousand, six hundred eighty-nine
4. Fifty-seven million, five hundred ninety thousand, one hundred twenty-three
5. Seventy-eight thousand, nine hundred fifty-four
6. Three hundred forty-five million, two hundred ninety-eight thousand, two hundred forty

7.	52,275	**8.**	125,305	**9.**	2,750	**10.**	2,305,482
11.	25,246,541	**12.**	382,406,400				
13.	546	**14.**	923	**15.**	9424	**16.**	3121
17.	6145	**18.**	1263	**19.**	1426	**20.**	11521
21.	1183	**22.**	851	**23.**	648	**24.**	85
25.	687	**26.**	2851	**27.**	1916	**28.**	682
29.	1162	**30.**	474	**31.**	$1215	**32.**	$669
33.	$299	**34.**	46	**35.**	4236	**36.**	2202 ft.
37.	2258 ft.	**38.**	6:35 PM	**39.**	11:07 PM	**40.**	11:00 AM
41.	1:50 PM	**42.**	1:47 PM	**43.**	3:55 PM		

Section B

1.	100958	**2.**	36225	**3.**	74368
4.	10710710	**5.**	24690000	**6.**	149509
7.	67275	**8.**	143724	**9.**	10080000
10.	123	**11.**	254	**12.**	524
13.	836	**14.**	305	**15.**	2004
16.	507	**17.**	705 rem 4	**18.**	804 rem 4
19.	$420 ft^2$	**20.**	$62730 ft^2$	**21.**	$165780
22.	$28	**23.**	$42	**24.**	$20
25.	$45	**26.**	$4416	**27.**	$2160
28.	85	**29.**	69	**30.**	85
31.	19	**32.**	22 ft., 11 in.	**33.**	47 ft., 8 in.
34.	412 quarts	**35.**	50 gal., 3 qt.	**36.**	4 lbs., 11 oz.

CHAPTER 2 : Integers, Exponents and Order of Operations

2A – Adding Integers

The **integers** are as follows: $\ldots, -3, -2, -1, 0, 1, 2, 3, \ldots$

The integers on the number line:
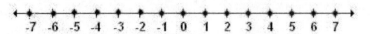

EXAMPLE 1: Adding on the number line.

A) $2 + 3 =$ **B)** $2 + (-3) =$ **C)** $-2 + (-3) =$ **D)** $-3 + 2 =$ **E)** $3 + (-2) =$

SOLUTION:

A)
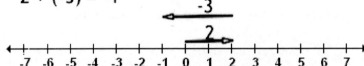

B)
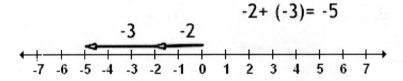

C)

$-2 + (-3) = -5$

D)

$-3 + 2 = -1$

E)
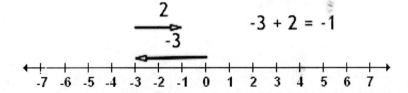

Definition: The **absolute value** of a non-negative number is the same as the number. The absolute value of a negative number is the number without its sign.

The absolute value of -2 is written as $|-2|$ and is equal to 2.

EXAMPLE 2: Evaluate the expression.

A) $|5|$ B) $|-6|$ C) $-|5|$ D) $-|-3|$

SOLUTION:

A) $|5| = 5$ B) $|-6| = 6$

C) $-|5| = -5$ C) $-|-3| = -3$

Rules for adding integers and signed numbers:

- **Adding two numbers with the same signs:**
  - Add their absolute values.
  - Use their common sign for the sign of the sum
- **Adding two numbers with different signs:**
  - Subtract the larger absolute value minus the smaller absolute value.
  - Use the sign of the number with the larger absolute value for the sum.

EXAMPLE 3: Add

A) $7 + 10$ B) $-30 + (-50)$ C) $7 + (-5)$

D) $-4 + 10$ E) $10 + (-12)$ F) $-20 + 15$

SOLUTION:

A) $7 + 10 = \mathbf{17}$

B) $-30 + (-50)$ Add their absolute values: $30 + 50 = 80$
                      Use their common sign for the sum: $-30 + (-50) = \mathbf{-80}$

C) $7 + (-5)$ Subtract the larger(in absolute value) minus the smaller: $7 - 5 = 2$
                    Use the sign of the number with larger absolute value: $7 + (-5) = \mathbf{2}$

D) $-4 + 10$ Subtract the larger (in absolute value) minus the smaller: $10 - 4 = 6$
                   Use the sign of the number with larger absolute value: $-4 + 10 = \mathbf{6}$

E) $10 + (-12)$ Subtract the larger(in absolute value) minus the smaller: $12 - 10 = 2$

Use the sign of the number with larger absolute value: $10 + (-12) = -2$

F) $-20 + 15$ Subtract the larger absolute value minus the smaller: $20 - 15 = 5$
Use the sign of the number with larger absolute value: $-20 + 15 = -5$

2A – EXERCISES

For $1 - 6$, add on the number line.

1. $2 + 4$

2. $2 + (-4)$

3. $-4 + 2$

4. $-3 + (-4)$

5. $-3 + 7$

6. $4 + (-7)$

For $7 - 24$, add.

7. $15 + 25$

8. $-15 + (-5)$

9. $-30 + (-20)$

10. $-45 + (-15)$

11. $-55 + (-5)$

12. $-32 + (-12)$

13. $15 + (-5)$

14. $25 + (-12)$

15. $35 + (-15)$

16. $12 + (-14)$

17. $25 + (-30)$

18. $24 + (-28)$

19. $-30 + 50$

20. $-25 + 75$

21. $-12 + 14$

22. $-15 + 5$

23. $-25 + 10$

24. $-42 + 15$

2A – WORKSHEET: Adding Integers

For 1 – 5, add on the number line.

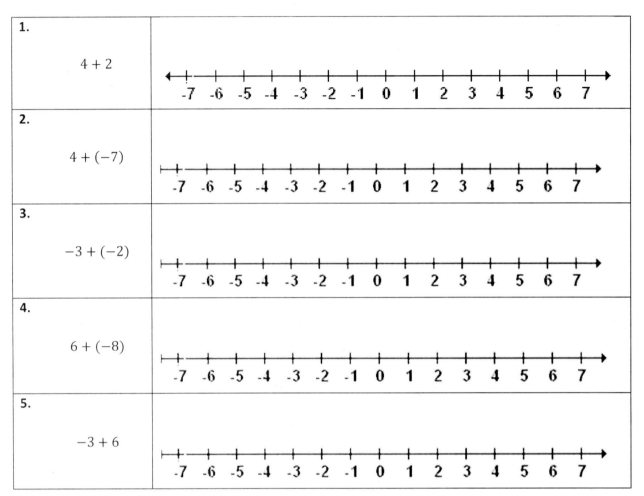

1.	$4 + 2$
2.	$4 + (-7)$
3.	$-3 + (-2)$
4.	$6 + (-8)$
5.	$-3 + 6$

For 6 - 23 , add.

6.	$25 + 72$	7.	$-32 + (-23)$	8.	$-15 + (-54)$
9.	$-36 + (-35)$	10.	$-73 + (-34)$	11.	$-65 + (-32)$

12. $70 + (-20)$	**13.** $15 + (-35)$	**14.** $60 + (-45)$
15. $43 + (-63)$	**16.** $34 + (-12)$	**17.** $45 + (-65)$
18. $-20 + 15$	**19.** $-15 + 40$	**20.** $-150 + 40$
21. $-80 + 100$	**22.** $-34 + 18$	**23.** $-45 + 90$

Answers:

1. 6	**2.** -3	**3.** -5	**4.** -2
5. 3	**6.** 97	**7.** -55	**8.** -69
9. -71	**10.** -107	**11.** -97	**12.** 50
13. -20	**14.** 15	**15.** -20	**16.** 22
17. -20	**18.** -5	**19.** 25	**20.** -110
21. 20	**22.** -16	**23.** 45	

2B – Subtracting Integers

Rule for subtracting integers: $a - b = a + (-b)$

EXAMPLE 1: Subtract

A) $5 - 10$ **B)** $-7 - 30$ **C)** $10 - (-5)$ **D)** $-15 - (-5)$

SOLUTION:

A) $5 - 10 = 5 + (-10) = -5$

B) $-7 - 30 = -7 + (-30) = -37$

C) $10 - (-5) = 10 + (-(-5)) = 10 + 5 = 15$
Notice that *when we subtract a negative we add a positive.*

D) $-15 - (-5) = -15 + 5 = -10$
Notice again that subtracting a negative is the same as adding a positive.

2B – EXERCISES

1. $25 - 15$ 2. $12 - 25$ 3. $17 - 28$

4. $15 - 30$ 5. $5 - 23$ 6. $8 - 17$

7. $3 - (-20)$ 8. $12 - (-13)$ 9. $14 - (-25)$

10. $28 - (-13)$ 11. $42 - (-26)$ 12. $33 - (-14)$

13. $25 - (-16)$ 14. $13 - (-32)$ 15. $15 - (-45)$

16. $-20 - 15$ 17. $-45 - 32$ 18. $-32 - 24$

19. $-18 - 17$ 20. $-23 - 15$ 21. $-41 - 28$

22. $-33 - 25$ 23. $-48 - 27$ 24. $-36 - 48$

25. $-12 - (-15)$ 26. $-25 - (-32)$ 27. $-15 - (-45)$

28. $-32 - (-75)$ 29. $-43 - (-12)$ 30. $-50 - (-15)$

31. $5 - 7 + 10 - 8$ 32. $-8 - 10 + 12 - 6$ 33. $3 - 15 + 20 - 5$

2B – WORKSHEET: Subtracting Integers

1. $30 - 15$	**2.** $13 - 25$	**3.** $15 - 28$
4. $20 - 30$	**5.** $7 - 23$	**6.** $5 - 17$
7. $10 - (-20)$	**8.** $15 - (-13)$	**9.** $12 - (-25)$
10. $25 - (-14)$	**11.** $43 - (-27)$	**12.** $35 - (-16)$
13. $30 - (-16)$	**14.** $17 - (-33)$	**15.** $20 - (-45)$
16. $-25 - 15$	**17.** $-47 - 32$	**18.** $-35 - 23$
19. $-19 - 14$	**20.** $-27 - 15$	**21.** $-45 - 26$
22. $-35 - 25$	**23.** $-58 - 27$	**24.** $-56 - 48$
25. $-12 - (-25)$	**26.** $-25 - (-42)$	**27.** $-15 - (-43)$
28. $-32 - (-15)$	**29.** $-43 - (-13)$	**30.** $-40 - (-15)$
31. $10 - 15 + 12 - 20$	**32.** $-8 - 5 - 10 + 4$	**33.** $-7 - 12 + 10 - 13$

Answers:

1.	15	**2.**	-12	**3.**	-13	**4.**	-10	**5.**	-16
6.	-12	**7.**	30	**8.**	28	**9.**	37	**10.**	39
11.	70	**12.**	51	**13.**	46	**14.**	50	**15.**	65
16.	-40	**17.**	-79	**18.**	-58	**19.**	-33	**20.**	-42
21.	-71	**22.**	-60	**23.**	-85	**24.**	-104	**25.**	13
26.	17	**27.**	28	**28.**	-17	**29.**	-30	**30.**	-25
31.	-13	**32.**	-19	**33.**	-22						

2C – Multiplying and Dividing Integers

Rules for multiplying integers and signed numbers:

- The product of two numbers with the same sign is a positive number.
- The product of two numbers with different signs is a negative number.

EXAMPLE 1:

A) $(-5)(12)$ **B)** $(7)(-6)$ **C)** $(-11)(-5)$ **D)** $(12)(6)$ **E)** $(3)(-2)(5)(-4)$

SOLUTION:

A) $(-5)(12) = -60$ **B)** $(7)(-6) = -42$

C) $(-11)(-5) = 55$ **D)** $(12)(6) = 72$ **E)** $(3)(-2)(5)(-4) = (-6)(5)(-4) = (-30)(-4) = 120$

Rules for dividing integers and signed numbers:

- The quotient of two numbers with the same sign is a positive number.
- The quotient of two numbers with different signs is a negative number.

EXAMPLE 2:

A) $\frac{25}{5}$ **B)** $\frac{-30}{6}$ **C)** $\frac{60}{-10}$ **D)** $\frac{-75}{-25}$

SOLUTION:

A) $\frac{25}{5} = 5$ **B)** $\frac{-30}{6} = -5$ **C)** $\frac{60}{-10} = -6$ **D)** $\frac{-75}{-25} = 3$

2C – EXERCISES:

1. $(-12)(7)$ 2. $(15)(-4)$ 3. $(30)(-7)$ 4. $(14)(-6)$

5. $(32)(-5)$ 6. $(7)(-15)$ 7. $(-20)(-5)$ 8. $(-45)(-23)$

9. $(-24)(-35)$ 10. $(-52)(-34)$ 11. $(46)(-32)$ 12. $(23)(-76)$

13. $\frac{-25}{5}$ 14. $\frac{36}{-18}$ 15. $\frac{-75}{15}$ 16. $\frac{60}{-12}$

17. $\frac{-10}{5}$ 18. $\frac{75}{-25}$ 19. $\frac{42}{-6}$ 20. $\frac{-56}{7}$

21. $\frac{-45}{-9}$ **22.** $\frac{-36}{-12}$ **23.** $\frac{-40}{-8}$ **24.** $\frac{-16}{-4}$

25. $\frac{-22}{-11}$ **26.** $\frac{-55}{-5}$ **27.** $\frac{-72}{-12}$ **28.** $\frac{-80}{-16}$

29. $(-5)(3)(-2)$ **30.** $(-4)(-3)(5)$ **31.** $(-7)(2)(-4)(-3)$

2C – WORKSHEET: Multiplying and Dividing Integers

1. $(-14)(8)$	**2.** $(18)(-5)$	**3.** $(37)(-8)$	**4.** $(24)(-6)$
5. $(42)(-7)$	**6.** $(8)(-25)$	**7.** $(-23)(-6)$	**8.** $(-55)(-33)$
9. $(-25)(-45)$	**10.** $(-62)(-32)$	**11.** $(56)(-33)$	**12.** $(43)(-75)$
13. $\dfrac{-35}{5}$	**14.** $\dfrac{54}{-18}$	**15.** $\dfrac{-45}{15}$	**16.** $\dfrac{60}{-4}$
17. $\dfrac{35}{-5}$	**18.** $\dfrac{125}{-25}$	**19.** $\dfrac{48}{-6}$	**20.** $\dfrac{-63}{7}$
21. $\dfrac{-54}{-9}$	**22.** $\dfrac{-48}{-12}$	**23.** $\dfrac{-56}{-8}$	**24.** $\dfrac{-64}{-4}$
25. $\dfrac{-33}{-11}$	**26.** $\dfrac{-77}{-7}$	**27.** $\dfrac{-84}{-12}$	**28.** $\dfrac{-96}{-16}$
29. $(5)(-2)(-3)$	**30.** $(-7)(2)(5)$	**31.** $(-3)(2)(-4)(5)$	**32.** $(-7)(-2)(-3)(5)$

Answers:

1. -112	**2.** -90	**3.** -296	**4.** -144	**5.** -294
6. -200	**7.** 138	**8.** 1815	**9.** 1125	**10.** 1984
11. -1848	**12.** -3225	**13.** -7	**14.** -3	**15.** -3
16. -15	**17.** -7	**18.** -5	**19.** -8	**20.** -9
21. 6	**22.** 4	**23.** 7	**24.** 16	**25.** 3
26. 11	**27.** 7	**28.** 6	**29.** 30	**30.** -70
31. 120	**32.** -210			

2D – Exponents and Order of Operations

The repeated multiplication, $2 \cdot 2 \cdot 2 \cdot 2 \cdot 2,$ can be expressed as:

$$2 \cdot 2 \cdot 2 \cdot 2 \cdot 2 = 2^5$$

In the expression $2^5,$ 2 is the **base**, and 5 is the **exponent**.

The expression 2^5 is read: "2 *to the* 5th *power*" or "2 *to the* 5th". **Power** is another word for exponent.

The expression 2^0 is equal to 1. In general, $a^0 = 1$, for all $a \neq 0$. (0^0 is undefined.)

EXAMPLE 1: Write using exponential notation.

A) $3 \cdot 3 \cdot 3 \cdot 3$ **B)** $5 \cdot 5 \cdot 5 \cdot 7 \cdot 7 \cdot 7 \cdot 7$ **C)** $(-2)(-2)(-2)(-2)$ **D)** 4

SOLUTION:

A) $3 \cdot 3 \cdot 3 \cdot 3 = 3^4$ **B)** $5 \cdot 5 \cdot 5 \cdot 7 \cdot 7 \cdot 7 \cdot 7 = 5^3 7^4$

C) $(-2)(-2)(-2)(-2) = (-2)^4$ **D)** 4^1

EXAMPLE 2: Write in expanded form.

A) 4^3 **B)** $2 \cdot 3^3$ **C)** $5^2 7^3$ **D)** $(-3)^3$

SOLUTION:

A) $4^3 = 4 \cdot 4 \cdot 4$ **B)** $2 \cdot 3^3 = 2 \cdot 3 \cdot 3 \cdot 3$ **C)** $5^2 7^3 = 5 \cdot 5 \cdot 7 \cdot 7 \cdot 7$ **D)** $(-3)^3 = (-3)(-3)(-3)$

EXAMPLE 3: Evaluate.

A) 2^3 **B)** $3 \cdot 2^2$ **C)** $(-2)^4$ **D)** $(-3)^3$ **E)** 5^0

SOLUTION:

A) 8 **B)** $3 \cdot 4 = 12$ **C)** 16 **D)** -27 **E)** 1

If an expression contains more than one operation, we do the operations in a specific order.

The Order of Operations:

- Evaluate expressions in parenthesis first.
- Evaluate expressions with exponents.
- Multiply or divide from left to right.
- Add or subtract from left to right.

EXAMPLE 4: Evaluate the mathematical expression.

1.	$5 + 3 \times 2$	**2.**	$2 \cdot 3^2$	**3.**	$8 - 2(3)$
4.	$6 - 2(5 - 3)$	**5.**	$2 - 5(4 - 8) \div 2$	**6.**	$(-1)3^2$
7.	-3^2	**8.**	$3(-2)^4$	**9.**	$\frac{10+4}{2}$
10.	$-2[3 + 2(5 - 8)] + 10$	**11.**	$10 - (5 - 2)^2$	**12.**	$3(2 - 5)^2 + 2(-5)$
13.	$2 - |3 - 7|$	**14.**	$-5|5 - 8| + 3$		

SOLUTION:

1. $5 + 3 \times 2 = 5 + 6 = 11$

2. $2 \cdot 3^2 = 2 \cdot 9 = 18$

3. $8 - 2(3) = 8 - 6 = 2$

4. $6 - 2(5 - 3) = 6 - 2(2) = 6 - 4 = 2$

5. $2 - 5(4 - 8) \div 2 = 2 - 5(-4) \div 2 = 2 + 20 \div 2 = 2 + 10 = 12$

6. $(-1)3^2 = (-1) \cdot 9 = -9$

7. $-3^2 = -9$ We can expand: $-3^2 = -3 \cdot 3 = -9$. Or, $-3^2 = (-1)3^2 = (-1)9 = -9$.

8. $3(-2)^4 = 3(16) = 48$

9. $\frac{10+4}{2} = \frac{14}{2} = 7$ Notice that the numerator (and denominator) groups an expression in the same way as parenthesis.

10. $-2[3 + 2(5 - 8)] + 10 = -2[3 + 2(-3)] + 10 = -2[3 - 6] + 10 = -2[-3] + 10 = 6 + 10 = 16$

11. $10 - (5 - 2)^2 = 10 - (3)^2 = 10 - 9 = 1$

12. $3(2 - 5)^2 + 2(-5) = 3(-3)^2 + 2(-5) = 3(9) + 2(-5) = 27 - 10 = 17$

13. $2 - |3 - 7| = 2 - |-4| = 2 - 4 = -2$ **14.** $-5|5 - 8| + 3 = -5|-3| + 3 = -5(3) + 3 = -12$

2D – EXERCISES

For 1 – 8, write using exponential notation.

1. $4 \cdot 4 \cdot 4$ **2.** $3 \cdot 5 \cdot 5 \cdot 5 \cdot 5$ **3.** $2 \cdot 2 \cdot 2 \cdot 7 \cdot 7$ **4.** $(-4)(-4)(-4)$

5. $-5 \cdot 5 \cdot 5 \cdot 5$ **6.** $(-5)(-5)(-5)(-5)$ **7.** $-2 \cdot 3 \cdot 3$ **8.** $-7 \cdot 7 \cdot 6 \cdot 6 \cdot 6$

For 9 - 16 , write in expanded form.

9. 3^5 **10.** $2 \cdot 5^3$ **11.** $2^3 7^4$ **12.** $-3 \cdot 5^2$

13. -3^5 **14.** $(-3)^5$ **15.** $-2^3 5^4$ **16.** $(-7)^3$

For 17 - 48 , evaluate.

17. $3 + 7 - 5$ **18.** $4 - 5 - 7 + 2$

19. $2 + 5 \times 3$ **20.** $(2 + 5) \times 3$

21. $\frac{10+2}{4+2}$ **22.** $10 - 2(3)$

23. $3 - 7(-4)$ **24.** $2 - 5(3 - 10)$

25. $6 + 5(2 - 4)$ **26.** $4 - 2(3 - 7) + 5$

27. $6 - 3(9 - 7)$ **28.** $3 + 4(5 - 8) - 6$

29. $-2[6 + 5(2 - 5)]$ **30.** 5^3

31. -5^2 **32.** $(-2)3^2$

33. $(-5)^2$ **34.** $(-2)^3$

35. -3^3 **36.** $(-3)^3$

37. $-3 \cdot 5^2$ **38.** -2^4

39. $3 + (5 - 2)^2$ **40.** $2 + 3(4 - 2)^3$

41. $5 + 2(3 - 5)^2$ **42.** $3^0 + 2(5 - 10)^2$

43. $3 - 2(3 - 5)^2$ **44.** $2 - 5(2 - 4)^3$

45. $15 - (5 - 3)^2 + 2(-3)$ **46.** $10 + (4 - 2)^3 + 3(-2)$

47. $3(-5) + (2 - 5)^3 + 3(-4)$ **48.** $15 - (2 - 7)^2 + 3(-5) + 3^0$

49. $-3|5 - 10|$

50. $3 - 4|2 - 5|$

51. $5 + 2|3 - 7|$

52. $10 - 2|4 - 8|$

2D – WORKSHEET: Exponents and Order of Operations

For 1 – 8, write using exponential notation.

1. $3 \cdot 3 \cdot 3$	**2.** $2 \cdot 7 \cdot 7 \cdot 7$	**3.** $4 \cdot 8 \cdot 8 \cdot 7 \cdot 7$	**4.** $(-3)(-3)(-3)$
5. $-7 \cdot 7 \cdot 7$	**6.** $(-4)(-4)(-4)$	**7.** $-5 \cdot 3 \cdot 3$	**8.** $-7 \cdot 7 \cdot 8 \cdot 8 \cdot 8$

For 9 - 16 , write in expanded form.

9. 7^5	**10.** $2 \cdot 7^3$	**11.** $4^3 8^4$	**12.** -9^2
13. -4^5	**14.** $(-4)^5$	**15.** $-2^3 8^4$	**16.** $(-6)^3$

For 17 - 48 , evaluate.

17. $4 + 10 - 5$	**18.** $8 - 2 - 7 + 2$
19. $2 + 7 \times 3$	**20.** $(2 + 7) \times 3$
21. $\dfrac{12+2}{4+3}$	**22.** $15 - 2(3)$
23. $3 - 5(-4)$	**24.** $8 - 5(2 - 10)$
25. $7 + 3(2 - 4)$	**26.** $9 - 2(4 - 7) + 6$

27. $8 - 2(9 - 7)$	**28.** $5 + 2(5 - 8) - 7$
29. $-2[7 + 4(3 - 5)]$	**30.** 4^3
31. -4^2	**32.** $(-2)4^2$
33. $(-4)^2$	**34.** $(-3)^3$
35. -4^3	**36.** $(-4)^3$
37. $-2 \cdot 5^2$	**38.** -3^4
39. $3 + (7 - 2)^2$	**40.** $4 + 3(5 - 2)^3$
41. $10 + 2(3 - 5)^2$	**42.** $4 + 2(6 - 10)^2$
43. $2^0 - 4(3 - 5)^2$	**44.** $5 - 3(2 - 4)^3$
45. $10 - (5 - 3)^2 + 2(-4)$	**46.** $15 + (4 - 2)^3 - 3(-2)$
47. $3(-4) + (2 - 5)^3 + 5(-4)$	**48.** $12 - (3 - 7)^2 + 3(-4) + 2^0$

49. $3 - 2\|7 - 5\|$	**50.** $-5\|7 - 10\| + 4$

Answers:

1.	3^3	**2.**	$2 \cdot 7^3$	**3.**	$4 \cdot 8^2 \cdot 7^2$
4.	$(-3)^3$	**5.**	-7^3	**6.**	$(-4)^3$
7.	$-5 \cdot 3^2$	**8.**	$-7^2 \cdot 8^3$	**9.**	$7 \cdot 7 \cdot 7 \cdot 7 \cdot 7$
10.	$2 \cdot 7 \cdot 7 \cdot 7$	**11.**	$4 \cdot 4 \cdot 4 \cdot 8 \cdot 8 \cdot 8 \cdot 8$	**12.**	$-9 \cdot 9$
13.	$-4 \cdot 4 \cdot 4 \cdot 4 \cdot 4$	**14.**	$(-4) \cdot (-4) \cdot (-4) \cdot (-4) \cdot (-4)$	**15.**	$-2 \cdot 2 \cdot 2 \cdot 8 \cdot 8 \cdot 8 \cdot 8$
16.	$(-6) \cdot (-6) \cdot (-6)$	**17.**	9	**18.**	1
19.	23	**20.**	27	**21.**	2
22.	9	**23.**	23	**24.**	48
25.	1	**26.**	21	**27.**	4
28.	-8	**29.**	2	**30.**	64
31.	-16	**32.**	-32	**33.**	16
34.	-27	**35.**	-64	**36.**	-64
37.	-50	**38.**	-81	**39.**	28
40.	85	**41.**	18	**42.**	36
43.	-15	**44.**	29	**45.**	-2
46.	29	**47.**	-59	**48.**	-15
49.	-1	**50.**	-11		

2E – Prime Factorization, Least Common Multiple and Greatest Common Divisor

The integer 3 is a **factor** of 6, since 3 divides 6 evenly (without a remainder). We can write 6 as the product $2 \cdot 3$ The word "factor" is also used as a verb, as in : we factor 6 as $2 \cdot 3$. We also say that $2 \cdot 3$ is a **factorization** of 6.

Note: Since a factor of a number divides into the number evenly, a factor is also called a **divisor**.

The number 12 can be factored as $1 \cdot 12$ or $3 \cdot 4$ or $2 \cdot 6$ or $2 \cdot 2 \cdot 3$. The numbers $1, 2, 3, 4, 6$ and 12 are all factors of 12.

An integer is called a **prime number** if it is greater than 1, and if its only factors are 1 and itself. The first five primes are: $2, 3, 5, 7$ and 11.

A **prime factorization** is a factorization where all the factors are prime numbers.

EXAMPLE 1:

A) List all the two-factor products of 18.　　　　**B)** Write 18 as a product of prime factors.

SOLUTION:

A) $1 \cdot 18, \ 2 \cdot 9, \ 6 \cdot 3$　　　　**B)** $2 \cdot 3 \cdot 3 = 2 \cdot 3^2$

The numbers 12 and 18 are both divisible by $1, 2, 3,$ and 6. The largest number that divides both is called the **greatest common divisor (GCD)**. Thus, 6 is the greatest common divisor of 12 and 18.

EXAMPLE 2: Find the GCD of 30 and 45.

SOLUTION:

The divisors of 30 are: $1, 2, 3, 5, 6, 10, 15,$ and 30.
The divisors of 45 are: $1, 3, 5, 9, 15,$ and 45.
The **GCD** is 15.

We can also find the GCD of two integers from their prime factorizations. The GCD contains the prime factors that are common to both numbers. Each prime factor appears the least number of times it occurs in each number.

EXAMPLE 3: Find the GCD of 18 and 60.

SOLUTION:

$18 = 2 \cdot 3 \cdot 3 = 2 \cdot 3^2$ and $60 = 2 \cdot 2 \cdot 3 \cdot 5 = 2^2 \cdot 3 \cdot 5$

The prime 2 appears once in 18 and twice in 60, so we include It once in the GCD. The prime 3 appears twice in 18 and once in 60, so we include it once in the GCD. The GCD is $2 \cdot 3 = 6$.

The number 6 is a **multiple** of 3 since $6 = 2 \cdot 3$. Stated differently, 3 divides 6 evenly. Some more multiples of 3 are 9, 12, 15, 18, 21, and 24.

The **least common multiple (LCM)** of two integers is the smallest positive integer that is a multiple of both integers.

EXAMPLE 4: Find the LCM of 12 and 18.

SOLUTION:

The first few multiples of 12 are: 12, 24, 36, 48.
The first few multiples of 18 are: 18, 36, 54, 72.
The number 36 is the smallest multiple that is common to both 12 and 18. It is the LCM.

We can also find the LCM of two integers from their prime factorizations. The LCM is the product of the prime factors of each number, each prime appearing the greatest number of times it occurs in each number.

EXAMPLE 5: Find the LCM of 36 and 120 from their prime factorizations.

SOLUTION:

The prime factorization of 36 is $2^2 3^2$.
The prime factorization of 120 is $2^3 3 \cdot 5$.
The LCM contains the prime factors: 2, 3, and 5. Prime number 2 occurs three times, prime number 3 occurs two times, and prime number 5 occurs one time.. The LCM is $2^3 3^2 5 = 360$.

2E – EXERCISES

For 1 – 4, state all the factors (divisors) of the given number.

1. 5   **2.** 9   **3.** 12   **4.** 42

5. What is a prime number?

For 6 – 9, give the prime factorization of the given number.

6. 3   **7.** 15   **8.** 21   **9.** 42

For 10 – 15, find the greatest common divisor of the pair of numbers.

10. 9 and 3   **11.** 15 and 10   **12.** 12 and 18

13. 75 and 30   **14.** 36 and 60   **15.** 125 and 75

For 16 – 21, find the least common multiple of the pair of numbers.

16. 5 and 7   **17.** 3 and 7   **18.** 7 and 14
19. 21 and 14   **20.** 30 and 42   **21.** 18 and 30

2E - WORKSHEET: Prime Factorization, Least Common Multiple and Greatest Common Divisor

For 1 – 6, find the prime factorization.

1. 24	2. 36	3. 28
4. 60	5. 45	6. 120

For 7 – 12, find the greatest common divisor.

7. 25 and 15	8. 12 and 10	9. 21 and 15
10. 24 and 18	11. 36 and 45	12. 60 and 45

For 13 – 18, find the least common multiple.

13. 12 and 6	14. 6 and 9	15. 10 and 15
16. 25 and 20	17. 12 and 15	18. 21 and 15

Answers:

1. $2^3 3$	2. $2^2 3^2$	3. $2^2 7$
4. $2^2 3 \cdot 5$	5. $3^2 5$	6. $2^3 3 \cdot 5$
7. 5	8. 2	9. 3
10. 6	11. 9	12. 15

13. 12   **14.** 18   **15.** 30

16. 100   **17.** 60   **18.** 105

2 – Answers to Exercises

Section A

1. 6	**2.** -2	**3.** -2	**4.** -7	**5.** 4
6. -3	**7.** 40	**8.** -20	**9.** -50	**10.** -60
11. -60	**12.** -44	**13.** 10	**14.** 13	**15.** 20
16. -2	**17.** -5	**18.** -4	**19.** 20	**20.** 50
21. 2	**22.** -10	**23.** -15	**24.** -27	

Section B

1. 10	**2.** -13	**3.** -11	**4.** -15	**5.** -18
6. -9	**7.** 23	**8.** 25	**9.** 39	**10.** 41
11. 68	**12.** 47	**13.** 41	**14.** 45	**15.** 60
16. -35	**17.** -77	**18.** -56	**19.** -35	**20.** -38
21. -69	**22.** -58	**23.** -75	**24.** -84	**25.** 3
26. 7	**27.** 30	**28.** 43	**29.** -31	**30.** -35
31. 0	**32.** -12	**33.** 3		

Section C

1. -84	**2.** -60	**3.** -210	**4.** -84
5. -160	**6.** -105	**7.** 100	**8.** 1035
9. 840	**10.** 1768	**11.** -1472	**12.** -1748
13. -5	**14.** -2	**15.** -5	**16.** -5
17. -2	**18.** -3	**19.** -7	**20.** -8
21. 5	**22.** 3	**23.** 5	**24.** 4
25. 2	**26.** 11	**27.** 6	**28.** 5
29. 30	**30.** 60	**31.** -168	

Section D

1. 4^3	**2.** $3 \cdot 5^4$	**3.** $2^3 \cdot 7^2$
4. $(-4)^3$	**5.** -5^4	**6.** $(-5)^4$
7. $-2 \cdot 3^2$	**8.** $-7^2 \cdot 6^3$	**9.** $3 \cdot 3 \cdot 3 \cdot 3 \cdot 3$
10. $2 \cdot 5 \cdot 5 \cdot 5$	**11.** $2 \cdot 2 \cdot 2 \cdot 7 \cdot 7 \cdot 7 \cdot 7$	**12.** $-3 \cdot 5 \cdot 5$
13. $-3 \cdot 3 \cdot 3 \cdot 3 \cdot 3$	**14.** $(-3)(-3)(-3)(-3)(-3)$	**15.** $-2 \cdot 2 \cdot 2 \cdot 5 \cdot 5 \cdot 5 \cdot 5$
16. $(-7)(-7)(-7)$		

17. 5	**18.** -6	**19.** 17	**20.** 21	**21.** 2	**22.** 4
23. 31	**24.** 37	**25.** -4	**26.** 17	**27.** 0	**28.** -15
29. 18	**30.** 125	**31.** -25	**32.** -18	**33.** 25	**34.** -8
35. -27	**36.** -27	**37.** -75	**38.** -16	**39.** 12	**40.** 26
41. 13	**42.** 51	**43.** -5	**44.** 42	**45.** 5	**46.** 12
47. -54	**48.** -24	**49.** -15	**50.** -9	**51.** 13	**52.** 2

Section E

1.	1, 5	**2.**	1, 3, 9	**3.**	1, 2, 3, 4, 6, 12	
4.	1, 2, 3, 6, 7, 14, 21, 42	**5.**	An integer greater than one, whose only divisors are one and itself.			

6.	3	**7.**	$3 \cdot 5$	**8.**	$3 \cdot 7$	**9.**	$2 \cdot 3 \cdot 7$
10.	3	**11.**	5	**12.**	6	**13.**	15
14.	12	**15.**	25	**16.**	35	**17.**	21
18.	14	**19.**	42	**20.**	210	**21.**	90

CHAPTER 3: Fractions

3A – Equivalent fractions; Mixed Numbers

A **fraction** is a part of a whole. In the diagram below we represent fractions as shaded areas.

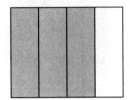

The rectangle on the left is divided into 4 equal parts. Three out of 4 of the parts are shaded. The shaded area illustrate the fraction $\frac{3}{4}$. The number 3 is the **numerator**, and 4 is the **denominator**.

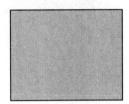 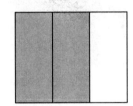

On the left, the entire first rectangle and 2 out of 3 (equal) parts of the second rectangle are shaded. The shaded areas illustrate the **mixed number** $1\frac{2}{3}$.

EXAMPLE 1: Write the fraction or mixed number illustrated by the shaded area(s).

A)

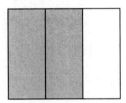

Solution: $\frac{2}{3}$

B)

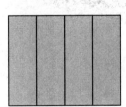

Solution: $\frac{4}{4} = 1$

C)

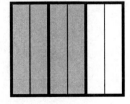

Solution: The diagram is divided into 6 equal parts, or into 3 equal parts (if we count 2 adjacent shaded areas as one part). We see that the shaded area is equal to $\frac{4}{6} = \frac{2}{3}$.

D)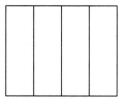

Solution: None of the rectangles are shaded. The fraction that represents the shaded area is $\frac{0}{4} = 0$.

E)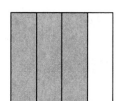

Solution: $1\frac{3}{4}$

Some special fractions:

1) $\frac{x}{x} = 1$ $(if\ x \neq 0)$. The same number over itself is equal to 1. (As in example 1B.)

2) $\frac{0}{x} = 0$ $(if\ x \neq 0)$. Zero over any number (except zero) is equal to 0. (As in example 1D.)

3) $\frac{x}{1} = x$. For example, $5 = \frac{5}{1}$.

Two fractions, $\frac{a}{b}$ and $\frac{c}{d}$ are **equivalent** (equal in value) if : $\frac{a \cdot x}{b \cdot x} = \frac{c}{d}$. In other words, if we multiply (or divide) both the numerator and denominator of a fraction by the same (non-zero) number, the result is an equivalent fraction. In example 1C we see that $\frac{4}{6} = \frac{2}{3}$, so, $\frac{4 \div 2}{6 \div 2} = \frac{2}{3}$.

EXAMPLE 2: Find the equivalent fraction with the given numerator or denominator.

A) $\frac{3}{4} = \frac{}{8}$

B) $\frac{2}{3} = \frac{4}{}$

C) $\frac{1}{2} = \frac{}{100}$

D) $\frac{3}{12} = \frac{}{4}$

SOLUTION:

A) $\frac{3 \cdot 2}{4 \cdot 2} = \frac{6}{8}$

B) $\frac{2 \cdot 2}{3 \cdot 2} = \frac{4}{6}$

C) $\frac{1 \cdot 50}{2 \cdot 50} = \frac{50}{100}$

D) $\frac{3 \div 3}{12 \div 3} = \frac{1}{4}$

Definition: Integer n contains **factor** m, if m divides n evenly.

For example, the number 6 contains factor 3, since 3 divides 6 evenly.

Definition: A fraction is expressed in **reduced form**, if the numerator and denominator do not share a common factor (other than 1). The process of finding the reduced form of a fraction is called "reducing the fraction to **lowest terms**."

EXAMPLE 3: Express the fraction in reduced form.

A) $\dfrac{4}{12}$

B) $\dfrac{10}{15}$

C) $\dfrac{9}{12}$

SOLUTION:

A) $\dfrac{4 \div 4}{12 \div 4} = \dfrac{1}{3}$

B) $\dfrac{10 \div 5}{15 \div 5} = \dfrac{2}{3}$

C) $\dfrac{9 \div 3}{12 \div 3} = \dfrac{3}{4}$

Below is an illustration of the mixed number $2\dfrac{3}{4}$:

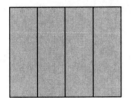 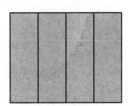 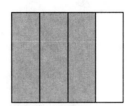

Notice the two wholes are divided into fourths. We see that $2\dfrac{3}{4}$ can also be written as $\dfrac{11}{4}$.

Every mixed number can be expressed as a fraction.

Writing a Mixed Number as a Fraction:

1) **Multiply the denominator times the whole.**
2) **Add the result to the numerator.**

EXAMPLE 4: Write as a fraction.

A) $5\dfrac{3}{4}$

B) $-4\dfrac{2}{3}$

C) 7

SOLUTION:

A) $5\dfrac{3}{4} = \dfrac{4 \times 5 + 3}{4} = \dfrac{23}{4}$

B) $-4\dfrac{2}{3} = -\dfrac{3 \times 4 + 2}{3} = -\dfrac{14}{3}$

C) $7 = \dfrac{7}{1}$

A fraction whose numerator is larger than its denominator is called an **improper fraction**. An improper fraction can be written as a mixed number.

Writing an Improper Fraction as a Mixed Number:

1) Divide the denominator into the numerator. Obtain a quotient and remainder.
2) The whole is the quotient and the fraction is the remainder over the denominator.

EXAMPLE 5: Write the improper fraction as a mixed number.

A) $\dfrac{5}{4}$

B) $\dfrac{8}{3}$

C) $\dfrac{23}{4}$

D) $-\dfrac{17}{3}$

SOLUTION:

A) $1\dfrac{1}{4}$

B) $2\dfrac{2}{3}$

C) $5\dfrac{3}{4}$

D) $-5\dfrac{2}{3}$

3A – EXERCISES

For 1 -4, write the fraction or mixed number represented by the shaded area(s).

1.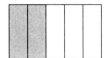

2.

3.

4.

For 5 - 9 , write in a simpler form.

5. $\dfrac{7}{7}$

6. $\dfrac{10}{5}$

7. $\dfrac{0}{3}$

8. $\dfrac{5}{1}$

9. $\dfrac{-3}{-3}$

For 10 - 15, find the equivalent fraction with the given numerator or denominator.

10. $\dfrac{3}{4} = \dfrac{}{12}$

11. $\dfrac{2}{5} = \dfrac{10}{}$

12. $\dfrac{7}{8} = \dfrac{}{16}$

13. $\dfrac{2}{3} = \dfrac{20}{}$ **14.** $\dfrac{4}{9} = \dfrac{}{27}$ **15.** $\dfrac{3}{4} = \dfrac{}{100}$

For 16 - 25, find the reduced form (reduce the fraction to lowest terms).

16. $\dfrac{6}{8}$ **17.** $\dfrac{12}{18}$ **18.** $\dfrac{10}{25}$ **19.** $-\dfrac{14}{21}$ **20.** $\dfrac{25}{30}$

21. $\dfrac{24}{36}$ **22.** $-\dfrac{2500}{10000}$ **23.** $\dfrac{360}{640}$ **24.** $\dfrac{300}{750}$ **25.** $\dfrac{350}{420}$

For 26 - 33, write the mixed number as an improper fraction.

26. $3\dfrac{4}{5}$ **27.** $7\dfrac{2}{3}$ **28.** $-4\dfrac{2}{7}$ **29.** $-5\dfrac{3}{8}$

30. $12\dfrac{23}{27}$ **31.** $24\dfrac{13}{25}$ **32.** $17\dfrac{34}{45}$ **33.** $-22\dfrac{35}{51}$

For 34 - 41, write the fraction as a mixed number.

34. $\dfrac{11}{5}$ **35.** $\dfrac{17}{3}$ **36.** $\dfrac{28}{13}$ **37.** $-\dfrac{7}{2}$

38. $\dfrac{61}{27}$ **39.** $\dfrac{102}{32}$ **40.** $\dfrac{652}{24}$ **41.** $\dfrac{342}{13}$

3A – WORKSHEET: Equivalent fractions; Mixed numbers.

For 1 -4, write the fraction or mixed number represented by the shaded area(s).

1.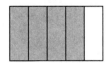

2.

3.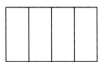

4.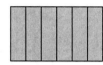

For 5 - 9 , write in a simpler form.

5. $\dfrac{9}{9}$	6. $\dfrac{12}{3}$	7. $\dfrac{0}{6}$	8. $\dfrac{8}{1}$	9. $\dfrac{-4}{-4}$

For 10 - 15, find the equivalent fraction with the given numerator or denominator.

10. $\dfrac{5}{6} = \dfrac{}{12}$	11. $\dfrac{2}{3} = \dfrac{10}{}$	12. $\dfrac{7}{8} = \dfrac{}{24}$
13. $\dfrac{2}{5} = \dfrac{20}{}$	14. $\dfrac{4}{9} = \dfrac{}{45}$	15. $\dfrac{3}{4} = \dfrac{}{32}$

For 16 - 25, find the reduced form (reduce the fraction to lowest terms).

16. $\dfrac{4}{12}$	17. $\dfrac{12}{15}$	18. $\dfrac{10}{35}$	19. $-\dfrac{12}{21}$	20. $\dfrac{25}{40}$
21. $\dfrac{25}{75}$	22. $-\dfrac{3200}{64000}$	23. $\dfrac{240}{720}$	24. $\dfrac{350}{750}$	25. $\dfrac{350}{490}$

For 26 -33 , write the mixed number as an improper fraction.

26. $6\dfrac{4}{5}$	27. $4\dfrac{2}{3}$	28. $-3\dfrac{2}{7}$	29. $-4\dfrac{3}{8}$
30. $15\dfrac{13}{25}$	31. $23\dfrac{32}{47}$	32. $24\dfrac{45}{57}$	33. $-14\dfrac{22}{35}$

For 34 - 41 , write the fraction as a mixed number.

34. $\frac{12}{5}$	35. $\frac{16}{3}$	36. $\frac{30}{13}$	37. $-\frac{9}{2}$
38. $\frac{62}{27}$	39. $\frac{105}{32}$	40. $\frac{753}{24}$	41. $\frac{452}{13}$

Answers:

1. $\frac{4}{5}$　　2. $1\frac{1}{2}$　　3. 0　　4. 1

5. $\frac{9}{9}=1$　6. $\frac{12}{3}=4$　7. $\frac{0}{6}=0$　8. $\frac{8}{1}=8$　9. $\frac{-4}{-4}=1$

10. $\frac{10}{12}$　　11. $\frac{10}{15}$　　12. $\frac{21}{24}$

13. $\frac{20}{50}$　　14. $\frac{20}{45}$　　15. $\frac{24}{32}$

16. $\frac{1}{3}$　17. $\frac{4}{5}$　18. $\frac{2}{7}$　19. $-\frac{4}{7}$　20. $\frac{5}{8}$

21. $\frac{1}{3}$　22. $-\frac{1}{20}$　23. $\frac{1}{3}$　24. $\frac{7}{15}$　25. $\frac{5}{7}$

26. $\frac{34}{5}$　　27. $\frac{14}{3}$　　28. $-\frac{23}{7}$　　29. $-\frac{35}{8}$

30. $\frac{388}{25}$　　31. $\frac{1113}{47}$　　32. $\frac{1413}{57}$　　33. $-\frac{512}{35}$

34. $2\frac{2}{5}$　　35. $5\frac{1}{3}$　　36. $2\frac{4}{13}$　　37. $-4\frac{1}{2}$

38. $2\frac{8}{27}$　　39. $3\frac{9}{32}$　　40. $31\frac{3}{8}$　　41. $34\frac{10}{13}$

3B – Multiplication and Division of Fractions

Rule for multiplying fractions: $\frac{a}{b} \cdot \frac{c}{d} = \frac{a \cdot b}{c \cdot d}$.

EXAMPLE 1: Find the product.

A) $\frac{2}{3} \cdot \frac{5}{7}$ **B)** $\frac{2}{3} \cdot \frac{3}{5}$ **C)** $\frac{5}{12} \cdot \frac{6}{11}$ **D)** $\frac{5}{6} \cdot \frac{3}{20}$

E) $\frac{4}{10} \cdot \frac{3}{2}$ **F)** $\left(2\frac{4}{5}\right)\left(1\frac{3}{7}\right)$ **G)** $\left(-2\frac{2}{5}\right)\left(\frac{3}{4}\right)$

SOLUTION:

A) $\frac{2}{3} \cdot \frac{5}{7} = \frac{2 \cdot 5}{3 \cdot 7} = \frac{10}{21}$

B)

$\frac{2}{3} \cdot \frac{3}{5} = \frac{2 \cdot 3}{3 \cdot 5} = \frac{2}{5}$

Divide numerator and denominator of the product by 3, so that the 3 from the numerator cancels with the 3 from the denominator.

C)

$\frac{5}{12} \cdot \frac{6}{11} = \frac{5}{2} \cdot \frac{1}{11} = \frac{5}{22}$

Divide the denominator of the first fraction and the numerator of the second by 6. This is the same as dividing the numerator and the denominator of the product by 6.

D) $\frac{5}{6} \cdot \frac{3}{20} = \frac{5}{2} \cdot \frac{1}{20} = \frac{1}{2} \cdot \frac{1}{4} = \frac{1}{8}$

Divide the denominator of the first fraction by 3 and the numerator of the second by 3. Divide the numerator of the first fraction by 5 and the denominator of the second by 5.

E) $\frac{4}{10} \cdot \frac{3}{2} = \frac{2}{5} \cdot \frac{3}{2} = \frac{3}{5}$

Reduce the first fraction by dividing the numerator and denominator by 2. Then divide the numerator of the first fraction by 2 and the denominator of the second fraction by 2.

F) $\left(2\frac{4}{5}\right)\left(1\frac{3}{7}\right) = \frac{14}{5} \cdot \frac{10}{7} = \frac{28}{7} = 4$

Change both mixed numbers to improper fractions.

G) $\left(-2\frac{2}{5}\right)\left(\frac{3}{4}\right) = -\frac{12}{5} \cdot \frac{3}{4} = -\frac{9}{5} = -1\frac{4}{5}$

Change the first mixed number to an improper fraction. Divide the numerator of the first fraction by 4 and the denominator of the second by 4. Write the product as a mixed number.

EXAMPLE 2:

A) In a college of 1430 students, $\frac{2}{5}$ are liberal arts majors. How many are liberal arts majors?

B) In a multilingual library, $\frac{2}{3}$ of the books are in English. There are a total of 726 books. How many are in English and how many are in another language?

C) A garden is $15\frac{3}{4}$ feet long, by $10\frac{2}{7}$ feet wide. Find the area of the garden in square feet.

SOLUTION:

A) $\frac{2}{5} \cdot \frac{1430}{1} = \frac{2}{1} \cdot \frac{286}{1} = 572$

B) $\frac{2}{3} \cdot \frac{726}{1} = \frac{2}{1} \cdot \frac{242}{1} = 484$ English books; $726 - 484 = 242$ non-English books

C) Recall the area of a rectangle: $Area = Length \times Width$.

$$Area = \left(15\frac{3}{4}\right)\left(10\frac{2}{7}\right) = \frac{63}{4} \cdot \frac{72}{7} = \frac{9}{1} \cdot \frac{18}{1} = \frac{162}{1} = 162 \text{ square feet.}$$

Definition: The **reciprocal** of fraction $\frac{a}{b}$ is fraction $\frac{b}{a}$.

For example, the reciprocal of $\frac{2}{3}$ is $\frac{3}{2}$.

Notice that the product of a fraction and its reciprocal is 1. For example, $\frac{2}{3} \cdot \frac{3}{2} = \frac{6}{6} = 1$.

We introduce division of fractions with the following division problem:

$$\frac{\frac{2}{3}}{\frac{5}{7}} \quad = \quad \frac{\frac{2}{3} \cdot \frac{7}{5}}{\frac{5}{7} \cdot \frac{7}{5}} \quad = \quad \frac{\frac{2}{3} \cdot \frac{7}{5}}{1} \quad = \quad \frac{2}{3} \cdot \frac{7}{5} \quad = \quad \frac{14}{15}$$

↑	↑	↑	↑
The divisor is $\frac{5}{7}$	Multiply numerator and denominator by $\frac{7}{5}$.	Notice that the denominator product is 1. $\frac{5}{7} \cdot \frac{7}{5} = 1$	*Notice that we actually multiply by the reciprocal of the divisor.*

Rule for division of fractions: $\frac{a}{b} \div \frac{c}{d} = \frac{a}{b} \times \frac{d}{c}$. That is, multiply the first fraction by the reciprocal of the second (multiply by the reciprocal of the divisor).

EXAMPLE 3: Find the quotient.

A) $\frac{2}{3} \div \frac{7}{8}$

B) $\frac{5}{7} \div \frac{10}{13}$

C) $3\frac{2}{5} \div \frac{7}{10}$

D) $\left(-2\frac{1}{5}\right) \div \left(-3\frac{2}{3}\right)$

SOLUTION:

A) $\dfrac{2}{3} \div \dfrac{7}{8} = \dfrac{2}{3} \cdot \dfrac{8}{7} = \dfrac{16}{21}$

B) $\dfrac{5}{7} \div \dfrac{10}{13} = \dfrac{5}{7} \cdot \dfrac{13}{10} = \dfrac{1}{7} \cdot \dfrac{13}{2} = \dfrac{13}{14}$

C) $3\dfrac{2}{5} \div \dfrac{7}{10} = \dfrac{17}{5} \cdot \dfrac{10}{7} = \dfrac{17}{1} \cdot \dfrac{2}{7} = \dfrac{34}{7} = 4\dfrac{6}{7}$

D) $\left(-2\dfrac{1}{5}\right) \div \left(-3\dfrac{2}{3}\right) = \dfrac{11}{5} \div \dfrac{11}{3} = \dfrac{11}{5} \cdot \dfrac{3}{11} = \dfrac{3}{5}$

3B – EXERCISES

For 1 – 12, find the product. Answer should be in reduced form.

1. $\dfrac{3}{4} \cdot \dfrac{5}{7}$

2. $\dfrac{2}{5} \cdot \dfrac{2}{3}$

3. $\dfrac{1}{3} \cdot \dfrac{3}{5}$

4. $\dfrac{4}{7} \cdot \dfrac{3}{8}$

5. $\dfrac{5}{7} \cdot \dfrac{4}{15}$

6. $(3)\left(\dfrac{2}{5}\right)$

7. $(5)\left(\dfrac{3}{5}\right)$

8. $\left(2\dfrac{1}{7}\right)\left(5\dfrac{1}{4}\right)$

9. $-\dfrac{3}{7} \cdot \dfrac{5}{12}$

10. $\left(-2\dfrac{5}{8}\right)\left(-4\dfrac{2}{7}\right)$

11. $\left(\dfrac{5}{6}\right)\left(\dfrac{3}{7}\right)\left(\dfrac{14}{25}\right)$

12. $\left(\dfrac{3}{5}\right)\left(\dfrac{2}{3}\right)\left(1\dfrac{1}{9}\right)$

13. In a college of 5487 students, $\dfrac{2}{3}$ are male. How many students are male?

14. In one day the postman delivered 375 packages. Exactly $\dfrac{3}{5}$ of the packages were from AMAZON. How many were not from AMAZON?

15. Kim processes insurance forms in a medical office. She has submitted $\dfrac{5}{8}$ of the 256 forms. How many remain?

16. The rapid strep test has a $\dfrac{1}{5}$ false negative rate. If a doctor's office does 75 rapid strep tests in one week, how many false negatives should they expect?

17. A yard is $20\dfrac{3}{8}$ feet wide and $15\dfrac{3}{4}$ feet long. Find the area of the yard in square feet.

For 18 - 27, find the quotient.

18. $\dfrac{2}{5} \div \dfrac{7}{11}$

19. $\dfrac{2}{5} \div \dfrac{3}{7}$

20. $\dfrac{5}{7} \div \dfrac{11}{14}$

21. $\dfrac{7}{8} \div \dfrac{21}{24}$

22. $\left(3\dfrac{2}{5}\right) \div \left(2\dfrac{1}{2}\right)$

23. $\left(-2\dfrac{3}{4}\right) \div \left(3\dfrac{3}{4}\right)$

24. $\left(-9\dfrac{1}{3}\right) \div \left(-\dfrac{4}{15}\right)$

25. $\dfrac{3}{7} \div 5$

26. $2\dfrac{5}{8} \div \dfrac{7}{12}$

27. $4\dfrac{1}{3} \div 5\dfrac{1}{5}$

3B – WORKSHEET: Multiplication and Division of Fractions

For 1 – 12, find the product. Answer should be in reduced form.

1. $\frac{2}{5} \cdot \frac{3}{11}$	**2.** $\frac{3}{5} \cdot \frac{5}{6}$	**3.** $\frac{3}{5} \cdot \frac{8}{15}$	**4.** $\frac{7}{12} \cdot \frac{8}{21}$
5. $\frac{3}{7} \cdot \frac{28}{75}$	**6.** $(4)\left(\frac{2}{5}\right)$	**7.** $(5)\left(\frac{2}{5}\right)$	**8.** $\left(2\frac{5}{6}\right)\left(2\frac{2}{3}\right)$
9. $-\frac{2}{7} \cdot \frac{5}{12}$	**10.** $\left(-1\frac{5}{8}\right)\left(-4\frac{2}{7}\right)$	**11.** $\left(-\frac{5}{6}\right)\left(\frac{3}{7}\right)\left(\frac{21}{25}\right)$	**12.** $\left(\frac{3}{5}\right)\left(\frac{2}{3}\right)\left(3\frac{1}{3}\right)$

13.	In a college of 5475 students, $\frac{2}{5}$ are science majors. How many students are science majors?
14.	Ken's bakery bakes 522 loaves of bread of which $\frac{2}{3}$ are whole wheat. How many are whole wheat and how many are not whole wheat?
15.	A telemarketer makes 736 phone calls in one day. Only $\frac{3}{32}$ of the calls end in a sale. How many calls end in a sale?
16.	A factory produces 675 light bulbs in one day. Exactly $\frac{3}{25}$ of the bulbs are defective. How many are defective?
17.	A court yard is $25\frac{3}{5}$ feet wide and $13\frac{3}{4}$ feet long. Find the area of the court yard in square feet.

For 18 - 25 , find the quotient.

18. $\frac{2}{5} \div \frac{7}{13}$	**19.** $\frac{2}{5} \div \frac{5}{7}$	**20.** $\frac{5}{7} \div \frac{11}{21}$	**21.** $\frac{7}{8} \div \frac{35}{48}$
22. $\left(2\frac{4}{5}\right) \div \left(3\frac{1}{2}\right)$	**23.** $\left(-2\frac{1}{3}\right) \div \left(4\frac{3}{4}\right)$	**24.** $\left(-7\frac{1}{3}\right) \div \left(-\frac{2}{15}\right)$	**25.** $\frac{2}{7} \div 4$

Answers:

1. $\frac{6}{55}$ **2.** $\frac{1}{2}$ **3.** $\frac{8}{25}$ **4.** $\frac{2}{9}$

5. $\frac{4}{25}$ **6.** $1\frac{3}{5}$ **7.** 2 **8.** $7\frac{5}{9}$

9. $-\frac{5}{42}$ **10.** $6\frac{27}{28}$ **11.** $-\frac{3}{10}$ **12.** $1\frac{1}{3}$

13. 2190   **14.** 348 whole wheat; 174 not whole wheat   **15.** 69   **16.** 81   **17.** $352 \, ft^2$

18. $\frac{26}{35}$ **19.** $\frac{14}{25}$ **20.** $1\frac{4}{11}$ **21.** $1\frac{1}{5}$

22. $\frac{4}{5}$ **23.** $-\frac{28}{57}$ **24.** 55   **25.** $\frac{1}{14}$

3C – Addition and Subtraction of Fractions

In the diagram below we illustrate addition of fractions whose denominators are equal. Notice that we add the numerators.

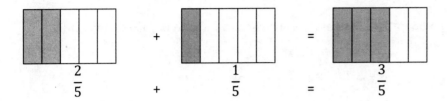

$$\frac{2}{5} \quad + \quad \frac{1}{5} \quad = \quad \frac{3}{5}$$

In the next diagram we illustrate addition of fractions whose denominators are different. Notice that we find equivalent fractions with a *common denominator* and add the numerators.

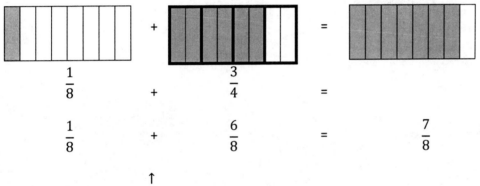

$$\frac{1}{8} \quad + \quad \frac{3}{4} \quad =$$

$$\frac{1}{8} \quad + \quad \frac{6}{8} \quad = \quad \frac{7}{8}$$

↑

Notice that 8 is a common denominator.

Fractions $\frac{5}{6}$ and $\frac{3}{10}$ have a common denominator of 60 since, $\frac{5}{6} = \frac{50}{60}$ and $\frac{3}{10} = \frac{18}{60}$. They also have a common denominator of 30 since, $\frac{5}{6} = \frac{25}{30}$ and $\frac{3}{10} = \frac{9}{30}$.

Definition: The smallest (positive) common denominator is called the **least common denominator**.

Finding the least common denominator:

- **Find the smallest multiple of the larger denominator containing the smaller denominator as a factor.**

For example, for $\frac{5}{6}$ and $\frac{3}{10}$, the larger denominator is 10. The first few multiples of 10 are: 10, 20, **30**, 40, ... The smallest multiple that contains factor 6 is 30, so 30 is the least common denominator.

Rule for adding or subtracting fractions:

- Express both fractions as equivalent fractions with the least common denominator. Add (or subtract) the numerators.

EXAMPLE 1: Perform the indicated operation. Express the answer as a fraction or mixed number (in lowest terms).

A) $\frac{5}{8} + \frac{1}{8}$ B) $\frac{5}{12} - \frac{1}{4}$ C) $\frac{2}{3} + \frac{4}{5}$ D) $\frac{3}{10} - \frac{1}{6}$ E) $2\frac{1}{3} + 3\frac{5}{6}$

SOLUTION:

A) $\frac{5}{8} + \frac{1}{8} = \frac{6}{8} = \frac{3}{4}$

B) $\frac{5}{12} - \frac{1}{4} = \frac{5}{12} - \frac{3}{12} = \frac{2}{12} = \frac{1}{6}$

C) $\frac{2}{3} + \frac{4}{5} = \frac{10}{15} + \frac{12}{15} = \frac{22}{15} = 1\frac{7}{15}$.

D) $\frac{3}{10} - \frac{1}{6} = \frac{9}{30} - \frac{5}{30} = \frac{4}{30} = \frac{2}{15}$

E) $2\frac{1}{3} + 3\frac{5}{6} = \frac{7}{3} + \frac{23}{6} = \frac{14}{6} + \frac{23}{6} = \frac{37}{6} = 6\frac{1}{6}$

EXAMPLE 2: Which fraction is larger?

A) $\frac{3}{4}$ or $\frac{5}{7}$ B) $\frac{4}{15}$ or $\frac{5}{21}$

SOLUTION:

A) Express both fractions with a common denominator.
$\frac{3}{4} = \frac{21}{28}$ and $\frac{5}{7} = \frac{20}{28}$ ← Since 21 is larger than 20, $\frac{3}{4}$ is larger than $\frac{5}{7}$.

B) Express both fractions with a common denominator. *The product of the two denominators is always a common denominator (however not necessarily the least common denominator).*

$\frac{4}{15} = \frac{4 \cdot 21}{15 \cdot 21} = \frac{84}{315}$ and $\frac{5}{21} = \frac{5 \cdot 15}{21 \cdot 15} = \frac{75}{315}$, since 84 is larger than 75, $\frac{4}{15}$ is larger than $\frac{5}{21}$.

EXAMPLE 3:

A) Sam painted $\frac{1}{2}$ of a room and Dan painted $\frac{1}{3}$ of the same room. How much of the room is painted?

B) A rectangular corral is $5\frac{2}{3}$ yards long and $3\frac{3}{4}$ yards wide. How much fencing is needed to enclose the corral?

C) Find the average of the three grades: $85, 72, 93$. Express the average as a mixed number.

SOLUTION:

A) $\frac{1}{2} + \frac{1}{3} = \frac{3}{6} + \frac{2}{6} = \frac{5}{6}$

B) $2\left(5\frac{2}{3}\right) + 2\left(3\frac{3}{4}\right) = \frac{2}{1} \cdot \frac{17}{3} + \frac{2}{1} \cdot \frac{15}{4} = \frac{34}{3} + \frac{30}{4} = \frac{136}{12} + \frac{90}{12} = \frac{226}{12} = 18\frac{10}{12} = 18\frac{5}{6}$ yards.

C) $\frac{85+72+93}{3} = \frac{250}{3} = 83\frac{1}{3}$

The area of a triangle is given by the following formula: $\boldsymbol{Area = \frac{1}{2}(base)(height)}$

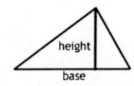

If the $height\ =\ 5\ in.$, and the $base\ =\ 10\ in.$,
$$Area = \frac{1}{2}(10)(5) = 25\ in^2$$

3C – EXERCISES

For 1 - 16, perform the indicated operation. Express your answer as a fraction or mixed number (in lowest terms).

1. $\frac{2}{53} + \frac{3}{53}$

2. $\frac{7}{13} + \frac{6}{13}$

3. $\frac{1}{4} + \frac{3}{8}$

4. $\frac{5}{12} - \frac{1}{6}$

5. $\frac{2}{5} - \frac{7}{10}$

6. $\frac{5}{12} + \frac{7}{18}$

7. $\frac{3}{4} - \frac{1}{6}$

8. $\frac{5}{21} + \frac{3}{14}$

9. $\frac{3}{10} - \frac{7}{15}$

10. $2\frac{1}{5} + 3\frac{2}{3}$

11. $5\frac{1}{3} - 2\frac{1}{4}$

12. $\frac{1}{5} + \frac{3}{4} + \frac{1}{2}$

13. $12\frac{3}{7} + 9\frac{5}{7}$

14. $46 - 13\frac{4}{7}$

15. $14\frac{3}{8} - 12$

16. $\frac{5}{12} + 34$

For 17 - 19, which fraction is larger?

17. $\frac{2}{3}$ or $\frac{11}{15}$

18. $\frac{5}{21}$ or $\frac{7}{15}$

19. $\frac{11}{50}$ or $\frac{6}{35}$

20. Karen ate $\frac{1}{4}$ of a pie, and Sam ate $\frac{2}{5}$ of the same pie. How much of the pie was eaten?

21. Joe bought $\frac{3}{4}$ of a yard of denim fabric, $2\frac{1}{2}$ yards of wool, and $\frac{3}{8}$ of a yard of silk. What is the total length of fabrics purchased?

22. Find the average of grades: $75, 83,$ and 95. Express the average as a mixed number.

23. A swimming pool is $4\frac{1}{2}$ yards long and $5\frac{2}{3}$ yards wide. Find the amount of fencing needed to enclose the swimming pool.

24. Find the largest fraction (compare two at a time): $\frac{4}{13}, \frac{1}{3}, \frac{7}{15}, \frac{4}{11}, \frac{2}{5}$

25. Find the smallest fraction (compare two at a time): $\frac{3}{4}, \frac{13}{20}, \frac{15}{19}, \frac{5}{8}, \frac{2}{3}$

26. Find the area of the triangle with base 10 inches and height 5 inches (Area of triangle $= \frac{1}{2}(base)(height)$).

27. Find the area of a triangle with base $\frac{2}{3}$ feet and height $\frac{1}{4}$ foot.

28. Find the area of the triangle with base $2\frac{1}{5}$ and height $3\frac{1}{2}$.

29. Find the area of the triangle with base 10 and height $5\frac{2}{5}$.

3C – WORKSHEET: Addition and Subtraction of Fractions

For 1 - 16, perform the indicated operation. Express your answer as a fraction or mixed number (in lowest terms).

1. $\frac{2}{79}+\frac{3}{79}$	2. $\frac{7}{17}+\frac{6}{17}$	3. $\frac{3}{4}+\frac{5}{8}$
4. $\frac{7}{12}-\frac{5}{6}$	5. $\frac{3}{5}-\frac{9}{10}$	6. $\frac{7}{12}+\frac{5}{18}$
7. $\frac{5}{9}-\frac{1}{6}$	8. $\frac{11}{21}+\frac{5}{14}$	9. $\frac{9}{10}-\frac{4}{15}$
10. $4\frac{2}{5}+2\frac{2}{3}$	11. $4\frac{1}{3}-3\frac{3}{4}$	12. $\frac{2}{7}+\frac{3}{4}+\frac{1}{2}$
13. $15\frac{5}{9}+23\frac{4}{9}$	14. $25-10\frac{2}{3}$	15. $13\frac{2}{5}-10$
16. $35-\frac{2}{3}$		

For 17 - 19, which fraction is larger?

17. $\frac{2}{3}$ or $\frac{7}{9}$	18. $\frac{5}{21}$ or $\frac{7}{19}$	19. $\frac{10}{55}$ or $\frac{4}{15}$

20.	Bob finished $\frac{1}{3}$ of his paper work on Monday and he finished another $\frac{1}{5}$ on Tuesday. How much of his paper work did he finish?

21.	Sue bought $\frac{5}{8}$ of a yard of denim fabric, $3\frac{1}{2}$ yards of wool, and $\frac{3}{4}$ of a yard of silk. What was the total length of fabrics purchased?
22.	Find the average of grades: $78, 89,$ and 92. Express the average as a mixed number.
23.	A garden is $3\frac{1}{2}$ yards long and $7\frac{2}{3}$ yards wide. Find the amount of fencing needed to enclose the garden.
24.	Find the largest fraction: $\frac{7}{12}, \frac{8}{13}, \frac{5}{8}, \frac{7}{10}, \frac{8}{15}$
25.	Find the smallest fraction: $\frac{1}{4}, \frac{6}{25}, \frac{7}{20}, \frac{4}{15}, \frac{7}{32}$

Answers:

1.	$\frac{5}{79}$	**2.**	$\frac{13}{17}$	**3.**	$1\frac{3}{8}$	**4.**	$-\frac{1}{4}$	**5.**	$-\frac{3}{10}$
6.	$\frac{31}{36}$	**7.**	$\frac{7}{18}$	**8.**	$\frac{37}{42}$	**9.**	$\frac{19}{30}$	**10.**	$7\frac{1}{15}$
11.	$\frac{7}{12}$	**12.**	$1\frac{15}{28}$	**13.**	39	**14.**	$14\frac{1}{3}$	**15.**	$3\frac{2}{5}$
16.	$34\frac{1}{3}$	**17.**	$\frac{7}{9}$	**18.**	$\frac{7}{19}$	**19.**	$\frac{4}{15}$	**20.**	$\frac{8}{15}$
21.	$4\frac{7}{8}$ yards	**22.**	$86\frac{1}{3}$	**23.**	$22\frac{1}{3}$ yards	**24.**	$\frac{7}{10}$	**25.**	$\frac{7}{32}$

3 - Answers to Exercises:

Section A:

1. $\frac{2}{5}$　　2. $1\frac{1}{2}$　　3. 0　　4. 1　　5. 1

6. 2　　7. 0　　8. 5　　9. 1　　10. $\frac{9}{12}$

11. $\frac{10}{25}$　　12. $\frac{14}{16}$　　13. $\frac{20}{30}$　　14. $\frac{12}{27}$　　15. $\frac{75}{100}$

16. $\frac{3}{4}$　　17. $\frac{2}{3}$　　18. $\frac{2}{5}$　　19. $-\frac{2}{3}$　　20. $\frac{5}{6}$

21. $\frac{2}{3}$　　22. $-\frac{1}{4}$　　23. $\frac{9}{16}$　　24. $\frac{2}{5}$　　25. $\frac{5}{6}$

26. $\frac{19}{5}$　　27. $\frac{23}{3}$　　28. $-\frac{30}{7}$　　29. $-\frac{43}{8}$　　30. $\frac{347}{27}$

31. $\frac{613}{25}$　　32. $\frac{799}{45}$　　33. $-\frac{1157}{51}$　　34. $2\frac{1}{5}$

35. $5\frac{2}{3}$　　36. $2\frac{2}{13}$　　37. $-3\frac{1}{2}$　　38. $2\frac{7}{27}$　　39. $3\frac{3}{16}$

40. $27\frac{1}{6}$　　41. $26\frac{4}{13}$

Section B:

1. $\frac{15}{28}$　　2. $\frac{4}{15}$　　3. $\frac{1}{5}$　　4. $\frac{3}{14}$　　5. $\frac{4}{21}$

6. $1\frac{1}{5}$　　7. 3　　8. $11\frac{1}{4}$　　9. $-\frac{5}{28}$　　10. $11\frac{1}{4}$

11. $\frac{1}{5}$　　12. $\frac{4}{9}$　　13. 3658　　14. 150　　15. 96

16. 15　　17. $320\frac{29}{32}\ ft^2$　　18. $\frac{22}{35}$　　19. $\frac{14}{15}$　　20. $\frac{10}{11}$

21. 1　　22. $1\frac{9}{25}$　　23. $-\frac{11}{15}$　　24. 35　　25. $\frac{3}{35}$

26. $4\frac{1}{2}$　　27. $\frac{5}{6}$

Section C:

1. $\frac{5}{53}$　　2. 1　　3. $\frac{5}{8}$　　4. $\frac{1}{4}$　　5. $-\frac{3}{10}$

6. $\frac{29}{36}$　　7. $\frac{7}{12}$　　8. $\frac{19}{42}$　　9. $-\frac{1}{6}$　　10. $5\frac{13}{15}$

11. $3\frac{1}{12}$　　12. $1\frac{9}{20}$　　13. $22\frac{1}{7}$　　14. $32\frac{3}{7}$　　15. $2\frac{3}{8}$

16. $34\frac{5}{12}$　　17. $\frac{11}{15}$　　18. $\frac{7}{15}$　　19. $\frac{11}{50}$

20. $\frac{13}{20}$　　21. $3\frac{5}{8}$ yards　　22. $84\frac{1}{3}$　　23. $20\frac{1}{3}$　　24. $\frac{7}{15}$

25. $\frac{5}{8}$　　26. $25\ in^2$　　27. $\frac{1}{12}ft^2$　　28. $3\frac{17}{20}$　　29. 27

Chapter 4 – Decimal Numbers

4A – Introduction to Decimal Numbers

The decimal number 25.397 has whole number part 25, and decimal part 397.

Place-Value Chart:

Hundreds	Tens	Ones	Decimal Point	Tenths	Hundredths	Thousandths	Ten-Thousandths
100	10	1	.	$\dfrac{1}{10}$	$\dfrac{1}{100}$	$\dfrac{1}{1000}$	$\dfrac{1}{10000}$
	2	5	.	3	9	7	

EXAMPLE 1: Write the word name.

A) .007   **B)** -3.61 **C)** 123.0542   **D)** .0075

SOLUTION:

A) Seven thousandths
B) Negative three and sixty-one hundredths
C) One hundred twenty-three and five hundred forty-two ten-thousandths
D) Seventy-five ten-thousandths

EXAMPLE 2: Write the decimal number as a fraction or mixed number.

A) .02   **B)** 3.054   **C)** .0025

SOLUTION:

A)

$$.02 \quad = \quad \frac{2}{100}$$

↑ two decimal places   ↑ two zeros

The number of decimal places is the same as the number of zeros in the denominator.

B) $3.054 = 3\frac{54}{1000}$

C) $.0025 = \frac{25}{10000}$

EXAMPLE 3:

A) Put in increasing order: 1.02, .952, .9499   **B)** Put in increasing order: .099, .15, .09099
C) Find the largest number: .95, .906, 1.01, 1.1, .99

SOLUTION:

A)

```
1 . 0 2 0 0
  . 9 5 2 0   ←
  . 9 4 9 9
```

Write the decimal numbers one underneath the other, lining up the decimal points. Fill in with trailing zeros (to the right of the decimal point) so that the numbers line up on the right. Now compare the numbers.

The numbers in increasing order: .9499, .952, 1.02

B) The numbers in increasing order: **C)** The largest number is 1.1 .
.09099, .099, .15

EXAMPLE 4: Round to the nearest hundredth.

A) 3.047   **B)** 2.543   **C)** .3257   **D)** 2.597

SOLUTION:

A) 3.047 ←   The digit to the right of the hundredth place is greater than or equal to 5, so we increase the hundredth place by 1, and drop the digits to the right of the hundredth place.

3.047 is rounded to 3.05

B) 2.543 ←   The digit to the right of the hundredth place is less than 5, so we drop the digits to the right of the hundredth place and do not increase.

2.543 is rounded to 2.54

C) .3257 ←   The digit to the right of the hundredth place is greater than or equal to 5, so we increase the hundredth place by 1, and drop the digits to the right of the hundredth place.

.3257 is rounded to .33

D) 2.597 ←   The digit to the right of the hundredth place is greater than or equal to 5, so we increase the hundredth place by 1. The hundredth place is 9, and $9 + 1 = 10$. Place the 0 in the hundredth place and add the 1 to tenth place.

2.597 is rounded to 2.60

4A – EXERCISES

For 1 – 4, write the word name.

1. .003
3. 357.0238

2. −2.54
4. .0023

For 5 – 10, write the decimal number as a fraction or mixed number.

5. .7
8. 3.258

6. 2.005
9. −2.0035

7. −.04
10. .0324

For 11 – 12, put in increasing order.

11. 1.01 , .978, .9528

12. .097, .13, .0909

For 13 – 15, round to the nearest thousandth.

13. 5.38742

14. .0348

15. 2.0597

16. Find the largest number: 2.01, 2.5, 2.099, 2.9, 2.85

17. Find the smallest number: 1.05, .9556, 1.035, .999

4A – WORKSHEET: Introduction to Decimal Numbers

For 1 – 4, write the word name.

1. .008	**2.** −7.64
3. 293.0546	**4.** .0056

For 5 – 10, write the decimal number as a fraction or mixed number.

5. .4	**6.** 3.006	**7.** −.07
8. 5.327	**9.** −5.0056	**10.** .0728

For 11 – 12, put in increasing order.

11. 1.001 , .982, .9437	**12.** .098, .12, .09099

For 13 – 15, round to the nearest thousandth.

13. 7.32849	14. .0237	15. 4.0797

16.   Find the largest number: $1.07, .96, .995, 1.095, 1.2$

17.   Find the smallest number: $.96, .908, .99, 1.1, .965$

Answers:

1.   Eight thousandth
2.   Negative seven and sixty-four hundredth
3.   Two hundred and ninety-three and five hundred forty-six ten-thousandths
4.   Fifty-six ten-thousandths

5. $\frac{2}{5}$

6. $3\frac{3}{500}$

7. $-\frac{7}{100}$

8. $5\frac{327}{1000}$

9. $-5\frac{56}{10000}$

10. $\frac{91}{1250}$

11. $.9437, .982, 1.001$

12. $.09099, .098, .12$

13. 7.328

14. $.024$

15. 4.080

16. 1.2

17. $.908$

4B – Addition and Subtraction of Decimal Numbers

When adding (or subtracting) decimal numbers we line up the decimal points. For example, $3.025 + .21$ is written:

$$
\begin{array}{r@{\,.\,}l}
3 & 0\ 2\ 5 \\
 & 2\ 1\ 0 \\
\hline
3 & 2\ 3\ 5
\end{array}
\quad \leftarrow \text{Fill in with a 0.}
$$

EXAMPLE 1: Perform the indicated operation.

A) $23.5 + 5.432$ **B)** $3 - .02$ **C)** $-1.36 - 54.035$
D) $2.08 + 53.7 + 4.256$ **E)** $3.1 - 2.005$ **F)** $-5.2 - (-7.01)$

SOLUTION:

A) Add:

$$
\begin{array}{r@{\,.\,}l}
2\ 3 & 5\ 0\ 0 \\
5 & 4\ 3\ 2 \\
\hline
2\ 8 & 9\ 3\ 2
\end{array}
$$

B) Subtract:

$$
\begin{array}{r@{\,.\,}l}
\overset{2}{\cancel{3}} & \overset{9}{\cancel{0}}\ \overset{10}{\cancel{0}} \\
 & 0\ 2 \\
\hline
2 & 9\ 8
\end{array}
$$

C) $-1.36 - 54.035 = -1.36 + (-54.035)$

Add:

$$
\begin{array}{r@{\,.\,}l}
-\quad\ \ 1 & 3\ 6\ 0 \\
-\ 5\ 4 & 0\ 3\ 5 \\
\hline
-\ 5\ 5 & 3\ 9\ 5
\end{array}
$$

D) $2.08 + 53.7 + 4.256 = 60.036$

E) $3.1 - 2.005 = 3.1 + (-2.005) = 1.095$

F) $-5.2 - (-7.01) = -5.2 + 7.01 = 1.81$

EXAMPLE 2:

A) In one month Ron's bills were as follows: $273.15 credit card, $53.72 cell phone, $125.38 electric, and $1525 for rent. What was his total?

B) A can of tuna fish costs $2.58 in the supermarket, and $3.27 in the corner grocery. How much more does it cost in the corner grocery?

C) Dale bought groceries whose total cost was $24.78 . He gave the cashier a $50 dollar bill. What was his change?

SOLUTION:

A) $1977.25

B) $.69

C) $25.22

4B – EXERCISES

1. $53.7 + 2.84$

2. $3 + 26.47 + .058$

3. $250 + 7.32 + .048$

4. $5.48 - 2.005$

5. $54.75 - 28.026$

6. $43. - .07$

7. $35 - .003$

8. $3.25 + 53.486$

9. $65.123 - 42.17$

10. $15.3 - 7.05$

11. $47.03 - 5.231$

12. $3.2 + 5.38 - 2.5$

13. $4.8 - .023$

14. $54.7 - .023$

15. $3.4 - .015$

16. Jen's shopping cart contains items that cost $5.86, $3.45, and $2.59. Find the total.

17. In September Dan's utility bills were as follows: $105.53 for gas, $65.78 for water and 45.62 for electricity. Find the total.

18. Find the amount of fencing required to enclose a garden that is 17.5 feet wide and 28.25 feet long.

19. A 16 oz. bottle of olive oil sells for $12.95 in the health food store, and the same bottle sells for $11.89 in the grocery store. How much more do you pay if you buy the olive oil in the health food store?

20. $47.3 + .072 + 24 + .0038$

21. $67 + .086 - 32 + .0057$

22. $86 - 3.567$

23. $489 - 324.856$

4B – WORKSHEET: Addition and Subtraction of Decimal Numbers

1. $78.7 + 2.94$	**2.** $5 + 32.48 + .078$	**3.** $425 + 9.37 + .056$
4. $4.43 - 2.007$	**5.** $79.45 - 35.036$	**6.** $36. - .07$
7. $23 - .004$	**8.** $4.35 + 72.483$	**9.** $75.125 - 51.14$
10. $25.4 - 3.05$	**11.** $6.241 - 27.03$	**12.** $4.2 + 5.65 - 3.5$
13. $5.8 - .017$	**14.** $56.5 - .032$	**15.** $7.4 - .012$

16. Ken's shopping cart contains items that cost $9.84, $13.45, and $12.49. Find the total.
17. In December Max's utility bills were as follows: $123.73 for gas, $85.75 for water and $75.62 for electricity. Find the total.
18. Find the amount of fencing required to enclose a garden that is 32.5 feet wide and 27.25 feet long.
19. A 64 oz. bottle of detergent sells for $13.45 in the grocery store, and the same bottle sells for $11.99 at Walmart. How much more do you pay if you buy the detergent in the grocery store?

20. $52.3 + .052 + 33 + .0021$	**21.** $53 + .087 - 25 + .0063$
22. $72 - 2.986$	**23.** $512 - 352.253$

Answers:

1.	81.64	**2.**	37.558	**3.**	434.426	**4.**	2.423
5.	44.414	**6.**	35.93	**7.**	22.996	**8.**	76.833
9.	23.985	**10.**	22.35	**11.**	-20.789	**12.**	6.35
13.	5.783	**14.**	56.468	**15.**	7.388	**16.**	$35.78
17.	$285.10	**18.**	119.5 ft.	**19.**	$1.46	**20.**	85.3541
21.	28.0933	**22.**	69.014	**23.**	159.747		

4C – Multiplication of Decimal Numbers

We multiply **factors** and the result is the **product**:

$$a \quad \times \quad b \quad = \quad c$$
$$\uparrow \qquad\qquad \uparrow \qquad\qquad \uparrow$$
$$\text{factor} \qquad \text{factor} \qquad \text{product}$$

Rule for multiplying decimal numbers:

- Ignore the decimal points in the factors and multiply.
- Place the decimal point in the product so that the number of decimal places in the product is equal to the sum of the numbers of decimal places in the two factors.

EXAMPLE 1: Find the product.

A) $(.007)(3.23)$ **B)** $(5.02)(-23.6)$ **C)** $(-42)(-.28)$

SOLUTION:

A) Multiply:

```
    3. 2 3      ← 2 decimal places
     .0 0 7     ← 3 decimal places
  _____
  .0 2 2 6 1    ← The product has 2 + 3 = 5 decimal places. We added a zero onto
                   the left.
```

Notice that we do not line up the decimal places.

B) Multiply:

```
     − 2 3. 6      ← 1 decimal place
         5. 0 2    ← 2 decimal places
   _____
         4 7 2
       0 0 0
   1 1 8 0
  _____
 − 1 1 8 .4 7 2    ← 1 + 2 = 3 decimal places; recall that the product of a negative
                      and a positive is a negative.
```

C) Multiply:

```
                   0 decimal places
   − 4 2           ← 2 decimal places
   − .2 8          ←
  _____
   3 3 6
   8 4             ← 0 + 2 = 2 decimal places; recall that the product of two
  _____            negatives is a positive.
 1 1. 7 6
```

EXAMPLE 2:

A) If one floor tile sells for $2.82, what is the cost of 75 tiles?

B) The exchange rate for 1 Euro dollar is $1.27 U.S. dollars. If a hotel room rate is $157 Euros a night, what is the rate in U.S. dollars.

C) A table is 9.25 feet long. What is the length in inches (1 foot = 12 inches).

D) A plumber earns $36.75 an hour. How much does he earn in a 48 hour week?

E) Carpeting costs $9.75 a square yard. If a room is 10 yards wide and 15 yards long, what is the cost of carpeting the room?

SOLUTION:

A) $2.82 \times 75 = \$211.50$

B) $157 \times 1.27 = \$199.39$

C) $9.25 \times 12 = 111$ inches

D) $36.75 \times 48 = \$1764$

E) First find the size of the room in square yards: $10 \times 15 = 150$ square yards.
The cost of carpeting is: $150 \times 9.75 = \$1462.50$

The **circumference** of a circle is the distance around the edge of the circle. The **radius** is the distance from the center of the circle to the circumference.

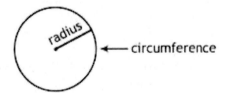

The ratio between the circumference and twice the radius is a decimal number with an infinite number of decimal places represented by the Greek letter π (pi) . **The value of π is approximately 3.14.**

The circumference of a circle with radius r is given by the formula: $C = 2\pi r$ (C = circumference).

The area of a circle of radius r is given by the formula: $A = \pi r^2$ (A = area).

EXAMPLE 3: For a circle of radius 5 inches, find the circumference and the area. Approximate π with the value 3.14.

SOLUTION:

$C = 2\pi r \cong 2(3.14)(5) = 31.4$ inches.   (The symbol, $\cong$, means "approximately equal to" .)

$A = \pi r^2 \cong (3.14)(5)^2 = (3.14)(25) = 78.5$ square inches.

Notice that the circumference is measured in inches and the area in square inches.

EXAMPLE 4: For a circle of radius 4.2 feet, find the circumference and the area. Do not approximate π.

SOLUTION:

$C = 2\pi r = (2\pi)(4.2) = 8.4\pi$ inches.

$A = \pi r^2 = \pi(4.2)^2 = 17.64\pi$ square inches.

A **power of 10** is a number that can be represented as 10^n , where n is an integer. The following are all powers of 10:

$$10, \ 10^2 = 100, 10^3 = 1000, \ 10^4 = 10000$$

Notice that (when n is positive) 10^n is equal to a 1 followed by n zeros.

Multiplication of a decimal number by 10^n (where n is positive) moves the decimal n places to the right.

EXAMPLE 5: Multiply

A) $(3.53)(10)$ **B)** $(2.58)(1000)$

SOLUTION:

A) $(3.53)(10) = 35.3$ **B)** $(2.58)(1000) = 2580$ ← We added a zero onto the end.

4C – EXERCISES

For 1 – 18, find the product.

1. $(1.53)(.04)$	**2.** $(.007)(3.52)$	**3.** $(.005)(73.4)$
4. $(1.02)(.057)$	**5.** $(3.04)(.00023)$	**6.** $(2.07)(.056)$
7. $(.0085)(2.3)$	**8.** $(.058)(.32)$	**9.** $(4.08)(.0042)$
10. $(.0017)(.028)$	**11.** $(5.04)(30.2)$	**12.** $(45.2)(1.03)$
13. $(4.56)(10)$	**14.** $(3.82)(100)$	**15.** $(.005)(100000)$

16. $(2.3)(5000)$ **17.** $(58.2)(3000)$ **18.** $(.78)(400)$

19. If one can of paint costs $5.95, how much do 12 cans cost?

20. The exchange rate for one Mexican Peso is $0.068 U.S. dollars (that is, 6.8 cents). If a cup of coffee costs 30 pesos, what is the cost in dollars?

21. The exchange rate for 1 Euro dollar is $1.27 U.S. dollars. If an airline ticket costs 307 Euros, what is the cost in dollars?

22. A table is 5.25 feet long. What is the length in inches?

23. Wall paper costs $5.23 a square yard. If a wall is 3 yards high and 4 yards long, how much will the wall paper cost?

24. Carpeting costs $6.25 a square yard. If a room is 5 yards long and 4 yards wide, how much does it cost to carpet the room?

25. An electrician earns $38.25 and hour. How much does he earn in a 42 hour week?

26. A barber charges $10.50 for a haircut. In one day he gives 28 haircuts. How much did he earn that day?

27. Find the area of a room that is 12.5 feet long and 10.2 feet wide.

28. Find the area and circumference of a circle of radius 12 (approximate $\pi \cong 3.14$).

29. Find the area and circumference of a circle of radius 7 (do not approximate π).

30. Find the area and circumference of a circle of radius 2 (approximate π by 3.14).

31. Find the area and circumference of a circle of radius 1.3 (do not approximate π).

32. Find the area and circumference of a circle of radius 4 (approximate π by 3.14).

33. Find the area and circumference of a circle of radius 3.2 (do not approximate π).

34. Sam bought 3 lbs. of grapes at $2.99 per pound, and 4 lbs. of apples at $1.89 per pound. What change will he get from a $20 bill?

35. Kim bought 2 loaves of bread at $2.35 a loaf and 3 cartons of milk at $2.19 a carton. How much change will she get from a $100 bill?

4C – WORKSHEET: Multiplication of Decimal Numbers

For 1 – 18, find the product.

1. $(2.35)(.06)$	**2.** $(.006)(7.23)$	**3.** $(.008)(82.3)$
4. $(2.04)(.046)$	**5.** $(2.07)(.00045)$	**6.** $(3.04)(.028)$
7. $(.0078)(3.4)$	**8.** $(.078)(.62)$	**9.** $(5.07)(.0095)$
10. $(.0019)(.043)$	**11.** $(7.03)(40.2)$	**12.** $(76.3)(2.09)$
13. $(5.27)(10)$	**14.** $(5.72)(100)$	**15.** $(.009)(100000)$
16. $(3.7)(5000)$	**17.** $(84.2)(3000)$	**18.** $(.73)(600)$

19.	If one loaf of bread costs $2.95, how much do 15 loaves cost?
20.	The exchange rate for one Mexican Peso is $0.068 U.S. dollars (that is, 6.8 cents). If a soft drink costs 20 pesos, what is the cost in dollars?
21.	The exchange rate for 1 Euro dollar is $1.27 U.S. dollars. If a hotel room costs 150 Euros, what is the cost in dollars?

22.	A table is 7.25 feet long. What is the length in inches?
23.	Flooring costs $15.23 a square yard. If a floor is 4 yards long and 3 yards wide what is the cost of covering the floor?
24.	Carpeting costs $5.20 a square yard. If a room is 6 yards long and 5 yards wide, how much does it cost to carpet the room?
25.	An electrician earns $43.75 and hour. How much does he earn in a 46 hour week?
26.	A repair man charges $25.75 an hour. In one day he works 12 hours. How much did he earn that day?
27.	Find the area of a room that is 20.5 feet long and 13.5 feet wide.
28.	Find the area and circumference of a circle of radius 13 (approximate $\pi \cong 3.14$).
29.	Find the area and circumference of a circle of radius 8 (do not approximate π).
30.	Find the area and circumference of a circle of radius 12 (do not approximate π).
31.	Tim bought 5 cans of soup at $1.59 a can and 3 cakes at $2.38 a cake. How much change will he get from a $20 bill?
32.	Gary bought 6 sodas for $1.65 a bottle and 5 loaves of bread for $2.25 a loaf. How much change will he get form a $100 bill?

Answers:

1.	.141	**2.**	.04338	**3.**	.6584	**4.**	.09384	**5.**	.0009315
6.	.08512	**7.**	.02652	**8.**	.04836	**9.**	.048165	**10.**	.0000817
11.	282.606	**12.**	159.467	**13.**	52.7	**14.**	572	**15.**	900
16.	18500	**17.**	252600	**18.**	438	**19.**	$44.25	**20.**	$1.36
21.	$190.50	**22.**	87 inches	**23.**	$182.76	**24.**	$156	**25.**	$2012.50
26.	$309	**27.**	276.75 ft.	**28.**	A=530.66 C=81.64	**29.**	A=64π C=16π	**30.**	A = 144π C = 24π
31.	$4.91	**32.**	$78.85						

4D – Division of Decimal Numbers

Recall: We divide the **divisor** into the **dividend** and the result is the **quotient**. For example, $12 \div 3 = 4$, where 12 is the dividend, 3 is the divisor, and 4 is the quotient.

Division of a decimal number by a whole number: Place the decimal point in the quotient directly above the decimal point in the dividend.

EXAMPLE 1: Divide $19.55 \div 23$

SOLUTION:

```
                 .              ← place decimal point in quotient
        23 | 19  .  55


                .8  5
        2  3 | 1  9  .5  5
             1  8  4  ↓
                1  1  5
                1  1  5
                        0
```

EXAMPLE 2: Divide $15.96 \div 34$, round the quotient to the nearest hundredth (round to two decimal places).

SOLUTION:

```
               .4  6  9
        3  4 | 1  5  .9  6  0   ←insert a zero to hold the place
             1  3  6  ↓
                2  3  6
                2  0  4
                   3  2  0
                   3  0  6
                      1  4
```

Now round to the nearest hundredth: $.469 \cong .47$

Division of a decimal number by a decimal number: Move the decimal point in the divisor to the right until it is a whole number. Move the decimal point in the dividend the same number of places that the point in the divisor was moved.

For example,

$$3 \ . \ 2 \ 3 \ \overline{\smash{\big)}5 \ . \ 6 \ 7 \ 8} \qquad \rightarrow \qquad 3 \ 2 \ 3 \ \overline{\smash{\big)}5 \ 6 \ 7 \ . \ 8}$$

Move both decimal places two places to the right.

EXAMPLE 3:

A) $240.5 \div .37$

B) Divide; round the quotient to the nearest hundredth.
$2.85 \div 3.6$

SOLUTION:

A) Move both decimal places two places to the left and divide: $24050 \div 37 = 650$

B) The first three decimal places of $28.5 \div 36$ are $.791$
Rounded to nearest hundredth: $.79$

EXAMPLE 4:

A) A 15 pound bag of flour costs $34.80 . What is the price of one pound?

B) A 16 ounce jar of honey costs $8.64 and a 12 ounce jar costs $6.24. Which is cheaper per unit ounce?

C) If 13 gallons of gasoline cost $31.85, then what is the cost of 5 gallons?

D) Find the average of the following grades: $65, 78, 89,$ and $91.$ (Write the answer as a decimal number).

E) On a map .15 inch represents one mile. If the map distance between two towns is 27 inches, what is the distance in miles?

SOLUTION:

A) $34.80 \div 15 = \$2.32$
B) We divide: $8.64 \div 16 = \$0.54$ and $6.24 \div 12 = \$0.52.$ The 12 ounce jar is cheaper per unit ounce.

C) The cost of one gallon is : $31.85 \div 13 = \$2.45$. The cost of 5 gallons is: $(2.45)(5) = \$12.25$.

D) The average is: $\dfrac{65+78+89+91}{4} = \dfrac{323}{4} = 80.75$

E) The distance in miles is $27 \div .15 = 180$ miles.

EXAMPLE 5:

A) Convert $\dfrac{3}{8}$ to a decimal.

B) Convert $\dfrac{2}{7}$ to a decimal number and round to the nearest hundredth.

SOLUTION:

A) $3 \div 8 = .375$

B) $2 \div 7 = 2.857 \dots$
rounded to nearest hundredth: 2.86

Recall that a power of 10 is of the form 10^n. The following are some of the powers of 10: $10, 100, 1000, 10000$.

Division by 10^n , where n is positive, moves the decimal place n places *to the left*.

EXAMPLE 6: Divide

A) $\dfrac{123.4}{10}$

B) $\dfrac{345.6}{100}$

C) $\dfrac{5}{100}$

D) $\dfrac{0.23}{1000}$

SOLUTION:

A) 12.34

B) 3.456

C) 0.05

D) 0.00023

4D – EXERCISES

For 1 – 9, divide.

1. $53.56 \div 52$
2. $47.6 \div 17$
3. $6.560 \div 32$
4. $47.25 \div 6.3$
5. $246.1 \div .23$
6. $19.278 \div .54$
7. $854.4 \div 2.4$
8. $148.4 \div .28$
9. $8303 \div 2.3$

For 10 – 15, divide; round quotient to nearest hundredth.

10. $58.32 \div 3.1$
11. $35.24 \div .48$
12. $85.76 \div 35$
13. $42.73 \div 1.5$
14. $59.78 \div 2.4$
15. $273.4 \div 4.2$

16. It costs \$63.50 to fill a 25 gallon tank of gasoline. What is the price of one gallon of gas?

17. An avocado costs \$.75 . How many avocados can you buy with \$4.50 .

18. On a map, .12 inch represents 1 mile. If the map distance between New York City and Albany is 18 inches, how far is it in miles?

19. A 5 ounce jar of olive oil costs $6.15 and a 12 ounce jar costs $12.60 . Which one is cheaper per unit ounce?

20. If 4 inches on a map represent 35 miles, how many miles are represented by 11 inches?

21. A cereal box contains 12.5 ounces of cereal. How many boxes can be filled from 150 ounces of cereal?

22. Find the average of the following 4 grades: $89, 78, 97, 85$ (write the answer as a decimal number).

For 23 – 28, divide.

23. $5.4 \div .006$	24. $.63 \div .009$	25. $2.7 \div .03$
26. $3.6 \div .06$	27. $4.2 \div .007$	28. $.45 \div .09$

For 29 – 31, convert the fraction to a decimal.

29. $\dfrac{5}{8}$ 30. $\dfrac{3}{40}$ 31. $\dfrac{7}{40}$

For 32 – 34, convert to a decimal number and round to the nearest hundredth.

32. $\dfrac{3}{7}$ 33. $\dfrac{5}{11}$ 34. $\dfrac{3}{14}$

For 35 - 38, divide by the power of 10.

35. $\dfrac{567.8}{10}$ 36. $456.7 \div 100$ 37. $\dfrac{2.3}{1000}$ 38. $0.5 \div 10000$

4D – WORKSHEET: Division of Decimal Numbers

For 1 – 9, divide.

1. $62.72 \div 7$	**2.** $14.868 \div 4.2$	**3.** $299.6 \div .28$
4. $885.8 \div .43$	**5.** $6604 \div 1.3$	**6.** $2.2578 \div .53$
7. $1074.5 \div 3.5$	**8.** $8.815 \div 4.3$	**9.** $972 \div .24$

For 10 – 15, divide; round quotient to nearest hundredth.

10. $5.92 \div 7$	**11.** $3.96 \div 2.3$	**12.** $9.73 \div 4.8$
13. $4.24 \div 3.6$	**14.** $7.2 \div .34$	**15.** $3.86 \div 5.5$

16. It costs $76.25 to fill a 25 gallon tank of gasoline. What is the price of one gallon of gas?
17. A grapefruit costs $.55 . How many grapefruits can you buy with $4.40 .

18.	On a map, .15 inch represents 1 mile. If the map distance between New York City and Philadelphia is 14.25 inches, how far is it in miles?
19.	A 5 ounce jar of almond butter costs $11.50 and a 12 ounce jar costs $28.80 . Which one is cheaper per unit ounce?
20.	If 5 inches on a map represent 6.25 miles, how many miles are represented by 11 inches?
21.	A cereal box contains 24.5 ounces of cereal. How many boxes can be filled from 147 ounces of cereal?
22.	Find the average of the following 4 grades: $83, 74, 98, 84$ (write the answer as a decimal number).

For 23 – 28, divide.

23. $2.4 \div .06$	**24.** $.64 \div .08$	**25.** $3.2 \div .04$
26. $2.7 \div .09$	**27.** $5.4 \div .009$	**28.** $.72 \div .08$

For 29 – 31, convert to a decimal number.

29. $\frac{7}{8}$	30. $\frac{13}{40}$	31. $\frac{3}{25}$

For 32 – 34, convert to a decimal number and round to the nearest thousandth.

32. $\frac{5}{7}$	33. $\frac{7}{13}$	34. $\frac{9}{14}$

For 35 – 38, divide by the power of 10.

35. $987.6 \div 10$	36. $659.23 \div 100$	37. $\frac{2.589}{100}$	38. $\frac{0.56}{1000}$

Answers:

1.	8.96	**2.**	3.54	**3.**	1070	**4.**	2060
5.	5080	**6.**	4.26	**7.**	307	**8.**	2.05
9.	4050	**10.**	.85	**11.**	1.72	**12.**	2.03
13.	1.18	**14.**	21.18	**15.**	.70	**16.**	$3.05
17.	8	**18.**	95 miles	**19.**	5 oz. at $2.30 at ounce	**20.**	13.75 miles
21.	6	**22.**	84.75	**23.**	40	**24.**	8
25.	80	**26.**	30	**27.**	600	**28.**	9
29.	.875	**30.**	.325	**31.**	.12	**32.**	.714
33.	.538	**34.**	.643	**35.**	98.76	**36.**	6.5923
37.	.02589	**38.**	0.00056				

Answers to Exercises:

Section A

1. Three thousandth
2. Negative two and fifty-four hundredths
3. Three hundred and fifty-seven and two hundred thirty-eight ten-thousandths
4. Twenty-three ten-thousandths

5. $\frac{7}{10}$
6. $2\frac{1}{200}$
7. $-\frac{1}{25}$
8. $3\frac{129}{500}$
9. $-2\frac{7}{2000}$

10. $\frac{81}{2500}$
11. .9528, .978, 1.01
12. .0909, .097, .13
13. 5.387
14. .035

15. 2.060
16. 2.9
17. .9556

Section B

1. 56.54
2. 29.528
3. 257.368
4. 3.475
5. 26.724
6. 42.93
7. 34.997
8. 56.736
9. 22.953
10. 8.25
11. 41.799
12. 6.08
13. 4.777
14. 54.677
15. 3.385
16. $11.90
17. $216.93
18. 91.5 ft.
19. $1.06
20. 71.3758
21. 35.0917
22. 82.433
23. 164.144

Section C

1. .0612
2. .02464
3. .367
4. .05814
5. .0006992
6. .11592
7. .01955
8. .01856
9. .017136
10. .0000476
11. 152.208
12. 46.556
13. 45.6
14. 382
15. 500
16. 11500
17. 174600
18. 312
19. $71.4
20. $2.04
21. $389.89
22. 63 in.
23. $62.76
24. $125
25. $1606.50
26. $294
27. $127.5 ft^2$
28. A=452.16 C=75.36
29. A=49π C=14π
30. A = 12.56 C = 12.56
31. A = 1.69π C = 2.6π
32. A = 50.24 C = 25.12
33. A = 10.24π C = 6.4π
34. $3.47
35. $88.73

Section D

1. 1.03
2. 2.8
3. .205
4. 7.5
5. 1070
6. 35.7
7. 356
8. 530
9. 3610
10. 18.81
11. 73.42
12. 2.45
13. 28.49
14. 24.91
15. 65.10
16. $2.54
17. 6
18. 150 miles
19. 12 oz
20. 96.25 miles
21. 12
22. 87.25
23. 900
24. 70
25. 90
26. 60
27. 600
28. 5
29. .625
30. .075
31. .175
32. .43
33. .45
34. .21
35. 56.78
36. 4.567
37. .0023
38. .00005

CHAPTER 5: Percentages

5A – Percent, Part and Whole

A percent is a fraction of 100. *Per cent* means hundredths or per hundred. For example,

$$35\% = \frac{35}{100} = 0.35 \qquad\qquad 100\% = \frac{100}{100} = 1 \qquad\qquad 50\% = \frac{50}{100} = 0.5$$

$$150\% = \frac{150}{100} = \frac{15}{10} = 1.5 \qquad 2\% = \frac{2}{100} = 0.02 \qquad\qquad 12.5\% = \frac{12.5}{100} = 0.125$$

Changing a Percent to a Decimal

- **Move the decimal point two places to the left.**
- **Remove the % symbol.**

EXAMPLE 1: Change the percent to a decimal.

A)　　35%　　B)　　2%　　C)　　125%　　D)　　12.5%　　E)　0.3%

SOLUTION:

A)　　0.35　　B)　　0.02　　C)　　1.25　　D)　　0.125　　E)　　0.003

Changing a Decimal to a Percent

- **Move the decimal two places to the right.**
- **Add the % symbol.**

EXAMPLE 2: Change the decimal to a percent.

A)　　0.45　　B)　　1.12　　C)　　0.034　　D)　　0.004

SOLUTION:

A)　　45%　　B)　　112%　　C)　　3.4%　　D)　　0.4%

We use the following formula to find the part of a number:

$$\textbf{\textit{Part}} = (\textbf{\textit{Percent in decimal form}}) \times \textbf{\textit{Whole}}$$

For example, 30% of 60 is $0.30 \times 60 = 18$. Here 60 is the whole, 0.30 is the percent in decimal form, and 18 is the part.

EXAMPLE 3: Find

A) 25% of 12　　**B)** 12.5% of 16　**C)** 150% of 300　**D)** 5% of $13.80　**E)** 0.1% of 1000

SOLUTION:

A) $0.25 \times 12 = 3$　　　　**B)** $0.125 \times 16 = 2$　　　**C)** $1.50 \times 300 = 450$
D) $0.05 \times 13.80 = \$0.69$　**E)** $0.001 \times 1000 = 1$

Note: To find 10% or a number, move the decimal point one place to the left. For example, 10% of 38.6 is 3.86.

EXAMPLE 4:

A) Jan bought a coat on sale for 25% off the original price. If the original price was $94.00, what was the sales price?
B) The nutrition facts on a box of crackers states that there are 120 calories in a serving and that 30% of the calories are from fat. How many calories are from fat?
C) A class has 80 students. If 15% drop the class how many students remain in the class?
D) A furniture store owner buys a couch for $300 and sells it after increasing the price by 60%. What is the retail price?
E) In a class of 24 students 12.5% are men. How many women are in the class?

SOLUTIONS:

A) $(94)(.25) = 23.5,\ 94 - 23.5 = \$70.50$ sales price
B) $(120)(.3) = 36$ calories from fat
C) $(80)(.15) = 12,\ 80 - 12 = 68$ students remain in the class
D) $(300)(.6) = 180,\ 300 + 180 = \$480$ retail price
E) $(24)(.125) = 3,\ 25 - 3 = 21$ women in the class

We use the following formula to find a percent:

$$Percent = \frac{Part}{Whole} \cdot 100\%$$

EXAMPLE 5: Find the percent.

A) 12 is what percent of 60?
B) Sam got 4 hits out of 10 times at bat. What percent of the number of times at bat did Sam get hits?
C) A final exam has 40 questions. One must answer 25 out of 40 correctly to pass. Write the passing grade as a percent.
D) Jack earns $4600 per month. His monthly rent is $2070. What percent of his income is rent?

SOLUTION:

A) $\frac{12}{60} = 0.20 = 20\%$

B) $\frac{4}{10} = 0.4 = 40\%$

C) $\frac{25}{40} = 0.625 = 62.5\%$

D) $\frac{2070}{4600} = 0.45 = 45\%$

We use the following formula to find the whole:

$$Whole = \frac{Part}{Percent\ in\ decimal\ form}$$

EXAMPLE 6:

A) If 40% of a number is 100, what is the number?
B) If 15% of a number is 9.6, what is the number?
C) Jan's rent is 30% of her monthly salary. If her rent is $1500, what is her monthly salary?

SOLUTION:

A) $\frac{100}{.40} = 250$

B) $\frac{9.6}{.15} = 64$

C) $\frac{1500}{.3} = \$5000$

5A – EXERCISES

For 1 – 6, write as a decimal.

1. 25%	**2.** 5%	**3.** 150%
4. 0.3%	**5.** 12.5%	**6.** 0.02%

For 7 - 12, write as a percent.

7. 0.75	**8.** 0.5	**9.** 0.2
10. 1	**11.** 1.5	**12.** 0.004

For 13 - 21, find the part of the whole.

13. 25% of 40	**14.** 67% of 100	**15.** 10% of 15.28
16. 10% of 1230	**17.** 5% of 24.8	**18.** 12% of 27
19. 0.5% of 600	**20.** 13% of 12.3	**21.** 150% of 64

For 22 - 29, find the percent.

22. 18 is what percent of 36 ?	**23.** 16 is what percent of 64?
24. 9 is what percent of 90?	**25.** 3 is what percent of 60?
26. 12.8 is what percent of 64?	**27.** 3.2 is what percent of 32?
28. 7.68 is what percent of 64?	**29.** 21 is what percent of 300?

For 30 - 35, find the number (whole).

30. 20% of a number is 6, what is the number?	**31.** 15% of a number is 12, what is the number?
32. 8% of a number is 2.8, what is the number?	**33.** 35% of a number is 6.3, what is the number?
34. 12% of a number is 3, what is the number?	**35.** 24% of a number is 8.4, what is the number?

36. The original price of a suit is $250.00 . If there is a "20% off" sale, then what will the sale price be?

37. A college has 1450 students, where 52% are male. How many female students are there?

38. Sara works in a department store. She gets a discount of 10% off anything she buys in the store. What would she have to pay for a camera that originally costs $43.70?

39. A television set is marked "35% off" the regular price of $239. Find the amount of the discount and sale price.

40. A boy bought a baseball card for $5 and sold it after increasing the price by 20%. What was the sales price of the card?

41. A store owner bought a couch for $250 and sold it after increasing the price by 25%. What was the sales price of the couch?

42. At the start of the semester there are 40 students in a class. After one month 15% of the students drop the class. How many students are left in the class?

43. A student took an exam with 50 questions. She answered 70% of the questions before running out of time. How many questions did she leave out?

44. A man bought 80 handbags to sell at a flea market. He sold 45% of the handbags. How many handbags were not sold?

45. Ron picked apples from his backyard tree. He collected 155 apples and found that 62 had spots. What percent had spots?

46. Gordon spends $1161.00 each month in rent. He earns $2580.00 a month. What percent of his salary goes towards rent?

47. On the day of a snow storm, 15 out of 40 of Mr. Gale's students showed up for class. What percent came to school? What percent were absent?

48. On a final exam of 25 questions the passing grade is 15 correct. What is the percent of the questions must be answered correctly to pass the exam?

49. Joe paid $5 sales tax on an $80 item. What is the sales tax rate (expressed as a percent) ?

50. A class has 50 students and 20 students receive a grade of B. What percent of the class receive a B?

51. In a class of 40 students, 24 students came to class during a snow storm. What percent did not come to class?

52. Ira gave the waiter a tip of $3.75 that was 15% of his dinner bill. How much was his dinner bill?

53. Ken's sales commission rate is 24%. How much must he sell in order to earn $144 in commissions?

54. Karen bought a blouse and paid $2.80 in sales tax. If the sales tax rate is 8%, how much did the blouse cost?

55. Complete the table:

Fraction	Decimal	Percent
$\frac{5}{8}$		
	.28	
		345%

5A – WORKSHEET: Percent, Part and Whole

For 1 – 6, write as a decimal.

1. 32%	2. 3%	3. 120%
4. 0.5%	5. 15.2%	6. 0.07%

For 7 - 12, write as a percent.

7. 0.25	8. 0.7	9. 0.6
10. 2	11. 1.25	12. 0.003

For 13 - 21, find the part of the whole.

13. 20% of 80	14. 32% of 100	15. 10% of 23.87
16. 10% of 670	17. 5% of 64.8	18. 13% of 24
19. 0.2% of 800	20. 17% of 21.3	21. 125% of 40

For 22 - 29, find the percent.

22. 35 is what percent of 70 ?	23. 13 is what percent of 52?
24. 3 is what percent of 30?	25. 6 is what percent of 120?
26. 28 is what percent of 80?	27. 2.3 is what percent of 23?

| 28. 4.5 is what percent of 25? | 29. 21 is what percent of 600? |

For 30 - 35, find the number (whole).

30. 20% of a number is 4, what is the number?	31. 15% of a number is 27, what is the number?
32. 8% of a number is 6, what is the number?	33. 35% of a number is 22.75, what is the number?
34. 12% of a number is 6, what is the number?	35. 24% of a number is 16.8, what is the number?

36. Karen bought a textbook for $128 and sold it at the end of the semester for 65% of the original price. What was the selling price?

37. A store owner bought a bicycle for $225 and sold if after increasing the price by 80%. What was the selling price?

38. The original price of a book is $16. What is the price during a "45% off" sale?

39. In a class of 75 students 15 receive a grade of A. What percent receive an A?

40. A school has 970 students and 679 are female. What percent are male?

41. A company of 120 employees offers an insurance plan. The participation rate in the plan is 30%. How many employees do not participate in the plan?

42.	If a class has 55 students and 11 drop the class, what percent drop the class? What percent remain in the class?
43.	Dale sold an apartment through an sales agent who earns a 2% commission. Dale paid the agent $3000. What did the apartment sell for?
44.	Sam bought pants and paid $2.80 in tax. If the tax rate is 8%, what were the price of the pants?

45. Complete the table:

Fraction	Decimal	Percent
$\frac{3}{8}$		
	.43	
		175%

Answers:

1.	.32		2.	.03		3.	1.2	
4.	.005		5.	.152		6.	.0007	
7.	25%		8.	70%		9.	60%	
10.	200%		11.	125%		12.	0.3%	
13.	16		14.	32		15.	2.387	
16.	67		17.	3.24		18.	3.12	
19.	1.6		20.	3.621		21.	50	
22.	50%		23.	25%		24.	10%	
25.	5%		26.	35%		27.	10%	
28.	18%		29.	3.5%		30.	20	
31.	180		32.	75		33.	65	
34.	50		35.	70		36.	$83.20	
37.	$405		38.	$8.80		39.	20%	
40.	30% male		41.	84 do not participate		42.	Drop: 20% Remain: 80%	
43.	$150,000		44.	$35		46.		

46.

Fraction	Decimal	Percent
3/8	.375	37.5%
43/100	.43	43%
1 3/4	1.75	175%

5 – Answers to Exercises

Section A

1.	.25	**2.**	.05	**3.**	1.5
4.	.003	**5.**	.125	**6.**	.0002
7.	75%	**8.**	50%	**9.**	20%
10.	100%	**11.**	150%	**12.**	0.4%
13.	10	**14.**	67	**15.**	1.528
16.	123	**17.**	1.24	**18.**	3.24
19.	3	**20.**	1.599	**21.**	96
22.	50%	**23.**	25%	**24.**	10%
25.	5%	**26.**	20%	**27.**	10%
28.	12%	**29.**	7%	**30.**	30
31.	80	**32.**	35	**33.**	18
34.	25	**35.**	35	**36.**	$200
37.	696	**38.**	$39.33	**39.**	$83.65, $155.35
40.	$6	**41.**	$312.50	**42.**	34
43.	15	**44.**	44	**45.**	40%
46.	45%	**47.**	Present: 37.5% Absent: 62.5%	**48.**	60%
49.	6.25%	**50.**	40%	**51.**	40%
52.	$25	**53.**	$600	**54.**	$35

55.

Fraction	Decimal	Percent
$\frac{5}{8}$	.625	62.5%
$\frac{7}{25}$	.28	28%
$3\frac{9}{20}$	3.45	345%

CHAPTER 6: Scientific Notation

6A – Scientific Notation

Recall that $10^1 = 10, 10^2 = 100, 10^3 = 1000, 10^4 = 10000,$ *etc.*

Now, $10^{-1} = \frac{1}{10}, 10^{-2} = \frac{1}{100}, 10^{-3} = \frac{1}{1000}, 10^{-4} = \frac{1}{10000},$ *etc.*

Recall that multiplication by 10^n, where n is positive, moves the decimal place n places to the right.

Now, multiplication by 10^{-n}, where n is positive, moves the decimal place n places *to the left*. This is the same as division by a power of 10.

If n is positive, the expression $x \times 10^n$ represents the number x with the decimal place moved n places to the right. For example, $123 \times 10^5 = 12300000$ and $1.23 \times 10^5 = 123000$.

If n is negative, the expression $x \times 10^{-n}$ represents the number x with the decimal place moved n places to the left. For example, $3.24 \times 10^{-5} = .0000324$.

EXAMPLE 1: Find the product.

A) $(2.36)10^2$ **B)** 4.578×10^5 **C)** 54×10^3 **D)** 345.6×10^{-2} **E)** 67.8×10^{-6}

SOLUTION:

A) 236 **B)** 457800 **C)** 54000 **D)** 3.456 **E)** .0000678

The expression $x \times 10^n$ is written in **scientific notation** if $|x|$ is between 1 and 10, where $|x|$ can equal 1, but must be less than 10. We will call x the **coefficient** and 10^n the **power of 10**. The following are in scientific notation:

$2.3 \times 10^4 = 23000$

$3.26 \times 10^{-4} = .000326$

EXAMPLES 2:

1. Change from scientific notation to decimal notation: **A)** 5.67×10^5 **B)** 7.89×10^{-7}

2. Change from decimal notation to scientific notation: **A)** 36420000000 **B)** −.00000478

SOLUTIONS:

1. A) 567000 **B)** .000000789

2. A) 3.642×10^{10} **B)** -4.78×10^{-6}

Notice that $10^3 \times 10^5 = (1000)(100000) = 100000000 = 10^8$.

We multiply powers of 10 by *adding the exponents*: $10^3 \times 10^5 = 10^{3+5} = 10^8$.

We can multiply two numbers written in scientific notation:

$$(3 \times 10^5)(2 \times 10^4) = (3 \cdot 2) \times (10^5 \cdot 10^4) = 6 \times 10^9$$

Notice that we multiply the coefficients and multiply the powers of 10.

If the absolute value of the product of the coefficients exceeds 10 then we adjust the product so that the coefficient is between 1 and 10 in absolute value. For example,

$$(3 \times 10^5)(5 \times 10^4) = 15 \times 10^9 = (1.5 \times 10) \times 10^9 = 1.5 \times 10^{10} \ .$$

EXAMPLE 3: Multiply and write the answer in scientific notation

a. $(3.2 \times 10^5)(1.3 \times 10^7) =$

b. $(7 \times 10^{-3})(3 \times 10^7) =$

c. $(5 \times 10^{-4})(9 \times 10^{-7}) =$

SOLUTION:

a. 4.16×10^{12}

b. $21 \times 10^4 = (2.1 \times 10) \times 10^4 = 2.1 \times 10^5$

c. $45 \times 10^{-11} = (4.5 \times 10) \times 10^{-11} = 4.5 \times 10^{-10}$

Notice that $\frac{10^5}{10^2} = \frac{100000}{100} = 1000 = 10^3$.

We divide powers of 10 by *subtracting the exponents*: $\frac{10^5}{10^2} = 10^{5-2} = 10^3$.

We can divide two numbers in scientific notation:

$\frac{6\times10^7}{3\times10^4} = 2 \times 10^3$, notice that we divide the coefficients and divide the powers of 10 (*by subtracting the exponents*).

If the absolute value of the quotient of the coefficients is not in between 1 and 10, then we adjust. For example,

$\frac{1.4\times10^8}{2\times10^3} = .7 \times 10^5 = (7 \times 10^{-1}) \times 10^5 = 7 \times 10^4$.

EXAMPLE 4: Divide and express the quotient in scientific notation

A) $\frac{1.5\times10^7}{5\times10^4} =$

B) $\frac{6\times10^{-5}}{1.2\times10^{-7}} =$

SOLUTION:

A) $.3 \times 10^3 = (3 \times 10^{-1}) \times 10^3 = 3 \times 10^2$

B) 5×10^2

EXAMPLE 5:   First multiply, then divide.

A) $\frac{(2.4\times10^{-2})(3.2\times10^8)}{2\times10^3}$

B) $\frac{(8\times10^5)(2\times10^2)}{4\times10^{12}}$

SOLUTION:

A) $\frac{(2.4\times10^{-2})(3.2\times10^8)}{2\times10^3} = \frac{7.68\times10^6}{2\times10^3} = 3.84 \times 10^3$

B) $\frac{(8\times10^5)(2\times10^2)}{4\times10^{12}} = \frac{16\times10^7}{4\times10^{12}} = 4 \times 10^{-5}$

6A - EXERCISES

For 1 - 6 , change from decimal notation to scientific notation.

1. -356.2 **2.** 32100000
3. $-.0000000789$ **4.** .000637
5. 743.67   **6.** .00654

For 7-12, change from scientific notation to decimal notation

7. 3.2×10^9 **8.** 1.23×10^7
9. 7.89×10^{-5} **10.** 3.45×10^{-7}
11. 5.4×10^{-8} **12.** 9.7×10^6

For 13 -18 , multiply and write the result in scientific notation.

13. $(2 \times 10^7)(4 \times 10^3)$

14. $(3.2 \times 10^7)(1.3 \times 10^{-8})$

15. $(4 \times 10^{-9})(8 \times 10^{14})$

16. $(3 \times 10^{-8})(7 \times 10^{-5})$

17. $(4.1 \times 10^{15})(5 \times 10^6)$

18. $(-7 \times 10^{-6})(-5 \times 10^{-4})$

For 19 -23 , divide and write the result in scientific notation.

19. $\dfrac{8 \times 10^7}{4 \times 10^4}$

20. $\dfrac{1.5 \times 10^3}{5 \times 10^{-4}}$

21. $\dfrac{5 \times 10^{-4}}{2.5 \times 10^{-7}}$

22. $\dfrac{2.4 \times 10^3}{1.2 \times 10^7}$

23. $\dfrac{4.5 \times 10^7}{-9 \times 10^{-4}}$

For 24 - 29 , first multiply, then divide.

24. $\dfrac{(3.2\times10^6)(4.6\times10^5)}{2\times10^3}$

25. $\dfrac{(2.1\times10^7)(1.8\times10^{-2})}{3\times10^{-3}}$

26. $\dfrac{(5\times10^{-4})(2\times10^{-3})}{2\times10^{-2}}$

27. $\dfrac{(3.3\times10^{-5})(4.2\times10^2)}{3\times10^{-8}}$

28. $\dfrac{(4.5\times10^7)(3\times10^{-2})}{5\times10^8}$

29. $\dfrac{(3.6\times10^{-3})(7.2\times10^8)}{6\times10^{10}}$

6A - Worksheet : Scientific Notation

For 1 – 4, change from decimal notation to scientific notation.

1. 4560000	**2.** 567.98
3. .0000987	**4.** −.00000065

For 5 - 8, change from scientific notation to decimal notation.

5. 5.6×10^7	**6.** 4.56×10^3
7. -3.26×10^{-5}	**8.** 7.8×10^{-6}

For 9-12, multiply and write the answer in scientific notation.

9. $(4.2 \times 10^7)(2 \times 10^3)$	**10.** $(1.3 \times 10^{-5})(3.1 \times 10^{-9})$
11. $(7 \times 10^5)(2 \times 10^9)$	**12.** $(2.5 \times 10^{-5})(5 \times 10^{-3})$

For 13 – 16, divide and write the answer in scientific notation.

13. $\dfrac{6 \times 10^5}{2 \times 10^7}$	**14.** $\dfrac{5.6 \times 10^{-9}}{8 \times 10^{-3}}$
15. $\dfrac{1.2 \times 10^5}{2.4 \times 10^{-6}}$	**16.** $\dfrac{3.5 \times 10^{-7}}{7 \times 10^3}$

For 17 - 22, first multiply, then divide.

17. $\dfrac{(3.4\times10^8)(5.6\times10^5)}{2\times10^3}$	**18.** $\dfrac{(2.7\times10^8)(1.8\times10^{-4})}{3\times10^{-3}}$
19. $\dfrac{(0.5\times10^{-2})(6\times10^{-7})}{2\times10^{-3}}$	**20.** $\dfrac{(6.3\times10^{-7})(4.2\times10^2)}{3\times10^{-10}}$
21. $\dfrac{(4.5\times10^8)(4\times10^{-3})}{5\times10^{12}}$	**22.** $\dfrac{(4.5\times10^{-2})(7.2\times10^8)}{9\times10^{12}}$

Answers: 1. 4.56×10^6 **2.** 5.6798×10^2 **3.** 9.87×10^{-5} **4.** -6.5×10^{-7} **5.** 56000000

6. 4560 **7.** $-.0000326$ **8.** $.0000078$ **9.** 8.4×10^{10} **10.** 4.03×10^{-14}

11. 1.4×10^{15} **12.** 1.25×10^{-7} **13.** 3×10^{-2} **14.** 7×10^{-7} **15.** 5×10^{10}

16. 5×10^{-11} **17.** 9.52×10^{10} **18.** 1.62×10^7 **19.** 1.5×10^{-6} **20.** 8.82×10^5

21. 3.6×10^{-7} **22.** 3.6×10^{-6}

6 – Answers to Exercises

Section A

1. -3.562×10^2
2. 3.21×10^7
3. -7.89×10^{-8}
4. 6.37×10^{-4}
5. 7.4367×10^2
6. 6.54×10^{-3}
7. 3200000000
8. 12300000
9. $.0000789$
10. $.000000345$
11. $.000000054$
12. 9700000
13. 8×10^{10}
14. 4.16×10^{-1}
15. 3.2×10^6
16. 2.1×10^{-12}
17. 2.05×10^{22}
18. 3.5×10^{-9}
19. 2×10^3
20. 3×10^6
21. 2×10^3
22. 2×10^{-4}
23. -5×10^{10}
24. 7.36×10^8
25. 1.26×10^8
26. 5×10^{-5}
27. 4.62×10^5
28. 2.7×10^{-3}
29. 4.32×10^{-5}

Appendix 1 – Practice for Final Exam

Practice Set - 1

1.	A garden is 25 feet wide and 30 feet long. How many feet of fencing are needed to enclose the garden.
2.	Find the average of the following numbers: $87, 79, 90,$ and 64.
3.	A school yard is 130 feet long and 250 feet wide. Find the area of the school yard.
4.	Multiple: 507×370
5.	In June 3478 air conditioners were sold in Home Depot. In July 5200 air conditioners were sold. How many more air conditioners were sold in July?
6.	Divide: $16238 \div 23$
7.	Find the cost of seeding a lawn that is 25 yards wide and 76 yards long if the cost of seed is $3.00 a square yard.
8.	Max started his homework assignment at 2:25 PM and finished at 5:10 PM. How long did it take to complete the assignment?
9.	335 inches is equal to : _____feet and _____ inches.
10.	Combine: $-50 - 30 + (100)$
11.	Simplify: $10 - (15 - 12)^3 + 7 \times 3$
12.	Simplify: $6 - 2(8 - 13)$
13.	$\dfrac{5}{6} + \dfrac{7}{12}$
14.	$\dfrac{2}{5} + \dfrac{3}{4}$
15.	$\dfrac{3}{5} \div \dfrac{9}{10}$
16.	$\dfrac{10}{13} \times \dfrac{3}{5} \times \dfrac{26}{21}$
17.	$8\dfrac{3}{4} \div 2\dfrac{3}{8}$
18.	Express $12\dfrac{3}{8}$ as an improper fraction.
19.	Express $\dfrac{282}{25}$ as a mixed number.
20.	Which is the biggest fraction? a) $\dfrac{3}{4}$ b) $\dfrac{7}{12}$ c) $\dfrac{8}{19}$ d) $\dfrac{6}{7}$ e) $\dfrac{5}{8}$
21.	$6\dfrac{1}{4} \times 2\dfrac{2}{5}$

22.	Of 375 students in a school, $\frac{3}{5}$ took a biology class. How many students took biology?
23.	$34 - 12\frac{2}{3}$
24.	$12\frac{5}{8} + 13\frac{7}{8}$
25.	Reduce to lowest terms: $\frac{240}{300}$.
26.	$1.23 + 13 + .0087 + .0232$
27.	Change $\frac{5}{7}$ to a fraction rounded to the nearest hundredth.
28.	Dan bought 5 cartons of milk for $2.35 each, and 3 packages of cookies for $1.25 each. How much change does he get from a $20 bill?
29.	Which decimal number is the smallest? a) .0045 b) .023 c) .00087 d) .0209 e) .0104
30.	Round 34.97623 to the nearest hundredth.
31.	If 12 inches on a map represents 5 miles, how many miles do 15 inches represent?
32.	$3.2 \div .004$
33.	$78 - 3.762$
34.	$(1.03)(.034)$
35.	The decimal .067 converted to a fraction of equal value is _____.
36.	If 20% of a number is 55, what is the number?
37.	Find 12% of 27.
38.	The original price of a coat is $128. It went on sale at 15% off. What is the sale price?
39.	63 is what percent of 84?
40.	In a class of 60 students 36 are men. What percent of the students are women?
41.	Using scientific notation simplify: $\frac{(2.4\times10^{15})(3.2\times10^{-2})}{2\times10^4}$
42.	Find the prime factorization of : 180
43.	Find the greatest common divisor and the least common multiple of : 180 and 120

Answers:

1. 110	2. 80	3. $32500\ ft^2$
4. 187590	5. 1722	6. 706
7. $5700	8. 2hr, 45min	9. 27ft 11in
10. 20	11. 4	12. 16
13. 1 5/12	14. 1 3/20	15. 2/3
16. 4/7	17. 3 13/19	18. 99/8
19. 11 7/25	20. D	21. 15
22. 225	23. 21 1/3	24. 26 ½
25. 4/5	26. 14.2619	27. .71
28. $4.50	29. C	30. 34.98
31. 6.25 m	32. 800	33. 74.238
34. .03502	35. 67/1000	36. 275
37. 3.24	38. $108.80	39. 75%
40. 40%	41. 3.84×10^9	42. $2^2 3^2 5$

43. gcd: $2^2 3 \cdot 5 = 60$
    lcm: $2^3 3^2 5 = 360$

Practice Set - 2

1.	$22635 \div 45$
2.	257×705
3.	In year 2013 the number of freshmen in a college was 5600. In year 2014 the number was 4986. How many more freshmen were there in year 2014?
4.	Evaluate: $16 - 3(5 - 8)$
5.	Evaluate: $3(17 - 19)^2 - 5 \times 2$
6.	Find the cost of carpeting a room that is 7 yards along and 5 yards wide if carpeting costs $13 a square yard.
7.	A flight departs at 2:18 PM and arrives at 5:07 PM . How long is the flight?
8.	$\frac{3}{7} + \frac{2}{5}$
9.	$13\frac{5}{8} + 12\frac{7}{8}$
10.	$25\frac{1}{3} - 12\frac{2}{3}$
11.	Change to a mixed number: $\frac{376}{33}$
12.	$\frac{5}{12} \times \frac{3}{7} \times \frac{4}{15}$
13.	$\left(2\frac{1}{7}\right)\left(4\frac{2}{3}\right)$
14.	Which fraction is the smallest: $\frac{5}{7}, \frac{4}{5}, \frac{11}{13}, \frac{13}{15}$?
15.	Express as an improper fraction: $27\frac{14}{15}$
16.	Reduce: $\frac{320}{480}$
17.	Change $\frac{7}{11}$ to a decimal and round to the nearest hundredth.
18.	Sam bought 5 muffins for $1.12 each and 4 bottles of orange juice for $3.59 each. What change will he get from a $50 bill.
19.	$3.72 + 35 + .0046 + 15.9$
20.	$.307 \times .05$
21.	$1.2768 \div 0.42$
22.	Find the average: 32, 54, 75, 21, 42
23.	A salesman uses $27\frac{1}{2}$ gallons of gas on a 440 mile trip. How many miles can he travel on one gallon of gas?
24.	$49 \div 2\frac{5}{8}$
25.	Find 12% of 45.
26.	48 is 15% of what number?
27.	In a class of 35 students 20% are biology majors. How many students are not biology majors?
28.	Find the sale price of a $640 computer that is on sale for 15% off.
29.	If 20 inches on a map represent 30 miles how many miles are represented by 24 inches?
30.	Using scientific notation, simplify: $\frac{(3.3\times10^7)(2.1\times10^{-2})}{3\times10^2}$
31.	Find the prime factorization of 90.
32.	Find the greatest common divisor and the least common multiple of 60 and 200.
33.	Find the area of a triangle that has base $4\frac{1}{2}$ and height $5\frac{1}{3}$.
34.	$\frac{7}{8} - \frac{2}{5}$

35.	Convert the decimal number to a fraction: 0.068
36.	Convert the decimal number to a percent: 0.052
37.	Convert to feet and remaining inches: 375 inches
38.	Convert to pounds and remaining ounces: 75 ounces
39.	Convert to gallons and remaining quarts: 35 quarts
40.	Convert to ounces: 12 pounds
41.	Convert to inches: 32 feet
42.	Find the circumference and area of a circle of radius 4 (approximate π as 3.14).
43.	Find the circumference and area of a circle of radius 1.2 (do not approximate π).
44.	Write fourteen million, sixty-seven thousand, five as a numeral
45.	Convert 0.068 to a fraction in reduced form.
46.	Convert 0.037 to a percent.

Answers:

1.	503	2.	181185	3.	614
4.	25	5.	2	6.	$455
7.	2 hours 49 minutes	8.	$\frac{29}{35}$	9.	$26\frac{1}{2}$
10.	$12\frac{2}{3}$	11.	$11\frac{13}{33}$	12.	$\frac{1}{21}$
13.	10	14.	$\frac{5}{7}$	15.	$\frac{419}{15}$
16.	$\frac{2}{3}$	17.	0.64	18.	$30.04
19.	54.6246	20.	0.01535	21.	3.04
22.	44.8	23.	16 miles	24.	$18\frac{2}{3}$
25.	5.4	26.	320	27.	28
28.	$544	29.	36 miles	30.	2.31×10^3
31.	$2 \cdot 3^2 \cdot 5$	32.	gcd: 20, lcm: 600	33.	12
34.	$\frac{19}{40}$	35.	$\frac{17}{250}$	36.	5.2%
37.	31 ft., 3 in.	38.	4 lbs. , 11 oz.	39.	8 gal., 3 qts.
40.	192 oz.	41.	384 in.	42.	C=25.12, A=50.24
43.	C=2.4π, A=1.44π	44.	14,067,005	45.	$\frac{17}{250}$
46.	3.7%				

CPSIA information can be obtained
at www.ICGtesting.com
Printed in the USA
FFOW01n0957180515
13471FF

9 781465 273451